Scientific Principles of
MALTING AND BREWING

Charles W. Bamforth
Department of Food Science and Technology
University of California, Davis

AMERICAN SOCIETY OF
Brewing Chemists

Library of Congress Control Number: 2006926325
ISBN-13: 978-1-881696-08-7
ISBN-10: 1-881696-08-1

© 2006 by the American Society of Brewing Chemists
Second printing 2010
Third printing 2012
Fourth printing 2013
Fifth printing 2014

All rights reserved.
No portion of this book may be reproduced in any form, including photocopy, microfilm, information storage and retrieval system, computer database, or software, or by any means, including electronic or mechanical, without written permission from the publisher.

Reference in this publication to a trademark, proprietary product, or company name is intended for explicit description only and does not imply approval or recommendation to the exclusion of others that may be suitable.

Printed in the United States of America on acid-free paper

American Society of Brewing Chemists
3340 Pilot Knob Road
St. Paul, Minnesota 55121, U.S.A.

Dedication

For two of my favorite people:
Rick Swantz and the late Don Dunkley

Preface

This book is based on a class I teach at the University of California, Davis. As such, it anticipates that the reader has a basic grasp of the essential sciences, especially chemistry and biochemistry. In case there is a need to refresh one's memory, the first appendix should help.

The class—and therefore the book—addresses the entire brewing scenario, from crops in the field to beer in the body. It is perforce selective insofar as such an enormous topic deserves a volume 10 times the size of this one. However, I endeavor to point the reader in the directions they can go to satisfy their additional curiosities.

I thank the staff at ASBC headquarters for their hard work in preparing this book for publication. I acknowledge Claudia Graham for her excellent artwork. Thanks to the Institute of Brewing and Distilling for allowing me to reproduce (as Appendix 1) my series on brewing chemistry and biochemistry that was published in *The Brewer International* in 2001. I am very grateful to John Grigsby, Roger Rouch, and Dave Thomas for their valuable comments on the manuscript. Thanks also to Cindy-Lou Dull and Mike Joyce for steering the project through.

Above all, as ever, I publicly acclaim the devotion of my wife, Diane. I have dragged her thousands of miles in pursuit of my dream, and I love her very much. *My a'th kar*. This book is in part dedicated to her late father, Don Dunkley, not a great beer drinker but certainly a great man. It is also dedicated to my buddy, Rick Swantz, a beer bon vivant and a helluva good guy. Time for a pitcher and some chips!

Contents

Chapter 1
Some Introductory Comments 1

Chapter 2
Basics of Malting and Brewing 3

Chapter 3
Beer Types 9

Chapter 4
The Quality and Wholesomeness of Beer 13

Chapter 5
Barley and Malting 21

Chapter 6
The Components of Barley and Their Degradation
During Malting and Mashing 45

Chapter 7
Production of Sweet Wort 59

Chapter 8
Water 77

Chapter 9
Hops 83

Chapter 10
Wort Boiling, Clarification, and Cooling; Sugars 95

Chapter 11
Yeast 101

Chapter 12
Brewery Fermentations 113

Chapter 13
Beer Flavor: Its Nature, Origins, and Control 121

Chapter 14
Downstream Processing: Cold Conditioning, Filtration, and Stabilization 135

Chapter 15
Haze Instability 141

Chapter 16
Flavor Instability 151

Chapter 17
Foam 161

Chapter 18
Gushing 167

Chapter 19
Light Instability 169

Chapter 20
Biological Instability 171

Chapter 21
Packaging 173

Chapter 22
Quality Control and Quality Assurance 177

Chapter 23
Environmental Impacts and Outputs 185

Appendix 1
Chemistry and Biochemistry for Brewers 189

Appendix 2
Fundamental Statistics for Brewers 233

Index 237

Scientific Principles of MALTING AND BREWING

1. Some Introductory Comments

The word *beer* comes from the Latin *Bibere* (to drink). Merriam-Webster's dictionary pronounces:

> Etymology: Middle English *ber*, from Old English *bEor*; akin to Old High German *bior*
> An alcoholic beverage usually made from malted cereal grain, flavored with hops, and brewed by slow fermentation.

Compared with most other alcoholic beverages, beer contains relatively low levels of alcohol. The average strength of beer (alcohol by volume (ABV) = ml of ethanol per 100 ml of beer) in the United States is 4.6%. Internationally, the highest average ABV is 5.1% and the lowest is 3.9%. The ABV of wines is typically 11–15%.

The alcohol content is important in many countries as its level determines the amount of excise duty (tax) that the brewer needs to pay.

Taxes account for approximately 27% of the cost of producing a barrel of beer in the United States. Estimates of other costs are malt (3.5%), adjuncts (1.5%), hops (0.2%), packaging materials (26%), other materials (1.8%), production (20%), and sales and marketing (20%).

Units

In the United States, it is still common for brewers to measure temperature on the Fahrenheit scale, but the majority of brewers across the world employ the Celsius (Centigrade) measure:

$$°\text{Fahrenheit} = (°\text{Celsius} \times 9/5) + 32$$

Some of the major temperature ranges involved in malting and brewing are as follows:

	°F	°C
Steeping and germination	57–65	14–18
Kilning	122–230	50–110
Glucanolytic stands	104–122	40–50
Starch conversion	140–158	60–70
Wort separation	162–169	72–76
Boiling	203–212[a]	95–100
Fermentation	43–77	6–25
Conditioning	30	−1
Pasteurization	144–169	62–76

[a] Depending on altitude.

Other parameters are also quantified using different units in different countries. It is the norm in most countries other than United States to use hectoliters and kilograms.

Volume
    1 US barrel = 1.1734 hectoliters
    1 US barrel = 31 US gallons
    1 hectoliter = 100 liters
    1 US gallon = 128 fluid ounces

Weight
    1 ton = 1,000 kilograms (kg)
    1 kg = 2.205 pounds (lb)
    1 ounce = 28.35 grams (g)

Specific Gravity
    °Plato = °Brix = % (w/w) cane sugar
    10°Plato = specific gravity of 1.040

Hops
    1 Zentner = 50 kg of hops

Yeast
    10 million cells/ml ≈ 0.3 kg/hl

Carbon Dioxide
    1 vol. CO_2 per vol. of beer = 1.98 g/liter

Flavor Components
    1 ppm = 1 part per million = 1 mg/liter
    1 ppb = 1 part per billion = 1 µg/liter
    1 ppt = 1 part per trillion = 1 ng/liter

Alcohol
    Alcohol content by volume (ABV) = 1.26 × alcohol content by weight

Further Reading

Bamforth, C. (2003) Beer: Tap into the Art and Science of Brewing, 2nd ed. Oxford University Press, New York.

Hornsey, I. S. (2003) A History of Beer and Brewing. Royal Society of Chemistry, Cambridge, UK.

2. Basics of Malting and Brewing

Yeast (*Saccharomyces* spp.) can grow on sugar anaerobically by fermenting the sugar to ethanol:

$$C_6H_{12}O_6 \rightarrow 2C_2H_5OH + 2CO_2$$

If the sugars are derived from grapes, then the end product (spent growth medium) is *wine*. If the sugars are from apples, the product is hard *cider*. If the sugars are from grain, the product is *beer*.

Some beers are made chiefly from wheat and others from sorghum, but barley is the predominant grain used in the majority of the world's beers (Fig. 1).

Barley retains its husk on threshing, and this husk traditionally forms the filter bed through which the liquid extract of sugars is separated in the brewery.

Fig. 1. Two-row barley growing in Victoria, Australia. (Courtesy of Paul Schwarz)

The starch in barley is enclosed in cell walls and proteins, which are stripped away during the malting process (a limited germination of the barley grains). This softens the grain and makes it more readily milled. Unpleasant grainy and astringent characters are removed during malting.

Malting

Malting starts with *steeping* of barley in water at 14–18°C for up to 48 hr until it reaches a moisture content of 42–46%. This is usually achieved in a multi-stage process (typically three stages), with the steeps interspersed with "air rests" that allow the barley to get some oxygen.

Raising the moisture content allows the grain to germinate, a process that usually takes 3 to 5 days at 16–20°C. During *germination*, the enzymes break down the cell walls and some of the protein in the starchy endosperm, which is the grain's food reserve, rendering the grain friable (readily crumbled and thus more easily milled). Amylases are produced during germination that are important for breaking down the starch during the mashing process in the brewery.

The visible signs of malting are a softening of the grain, the emergence of rootlets, and the growth of the acrospire (shoot) underneath the husk.

Progressively increasing the temperature during *kilning* arrests germination, and regimes with increasing temperatures between 50 and perhaps 110°C are used to allow drying to <5% moisture content while preserving heat-sensitive enzymes. The more intense the kilning process, the darker the malt and the more roasted and burnt its flavor characteristics.

Brewing

In the brewery, the malted grain is first *milled* to produce relatively fine particles.

The particles are then intimately mixed with hot water in *mashing*. The water must possess the right mix of salts. For example, fine ales are historically produced from waters with high levels of calcium. Famous pilsners are from waters with low levels of calcium.

Typically, mashes consist of approximately three parts water to one part malt and incorporate a stand at about 65°C, at which temperature the granules of starch are converted by gelatinization from an indigestible crystalline state into a "melted" form that is much more susceptible to enzymatic digestion by amylases.

Some brewers add starch from other sources, such as maize or rice, to supplement that from malt. These other sources are called *adjuncts*.

After perhaps an hour of mashing, the liquid portion of the mash, known as *wort*, is recovered, either by straining through the residual *spent grains* or by filtering through plates.

The wort is run to the *kettle* (*copper*, because historically it was constructed from that metal) where it is boiled, usually for approximately 1 hr. *Boiling* serves various functions, including sterilization of the wort, precipitation of proteins (that would otherwise come out of solution in the finished beer and cause cloudiness), and the driving away of unpleasant grainy characters originating in the barley. Many brewers also add some

Fig. 2. Warrior hop cones. (Courtesy of Yakima Chief, Inc.)

adjunct sugars at this stage, at which most brewers introduce at least a proportion of their *hops* (Fig. 2).

The hops have two principal components of importance to the brewer: resins and essential oils.

The *resins* (so-called α-acids) are changed ("isomerized") during boiling to yield iso-α-acids, which provide the bitterness to beer. This process is rather inefficient. Nowadays, hops are often extracted at low temperatures with liquefied carbon dioxide and the extract is either added to the kettle or extensively isomerized outside the brewery for addition to the finished beer (thereby avoiding losses caused by the tendency of the bitter substances to stick to the yeast).

The *oils* are responsible for the "hoppy nose" on beer. They are very volatile; and if the hops are all added at the start of the boil, then all of the aroma will be blown away. In traditional lager brewing, a proportion of the hops is held back and added only toward the end of boiling, which allows a proportion of the aroma to remain in the wort (*late hopping*). In traditional ale production, a handful of hops is added to the cask at the end of the process, enabling a complex mixture of oils to give a distinctive character to such products. This is called *dry hopping*. Liquid carbon dioxide can be used to extract oils as well as resins, and these extracts can also be added late in the process to make modifications to beer flavor.

After the precipitate produced during boiling ("hot break," or trub) has been removed, the hopped wort is cooled, aerated, and *pitched* with yeast. The *cold break* that precipitates when wort is cooled may be removed prior to adding yeast, but not always.

Brewing yeast (Fig. 3) can be divided into ale (*Saccharomyces cerevisiae*) and lager (*S. pastorianus*) strains. *S. cerevisiae* collects on the surface of the fermenting wort, and *S. pastorianus* settles to the bottom of a fermentation. Both types need a little oxygen to trigger their metabolism, but otherwise the alcoholic *fermentation* is anaerobic.

Ale fermentations are usually complete within a few days at temperatures as high as 20°C, whereas lager fermentations at temperatures as low as 6°C can take several weeks.

Fermentation is complete when the desired alcohol content has been reached and when an unpleasant butterscotch flavor (caused by diacetyl) that develops during all

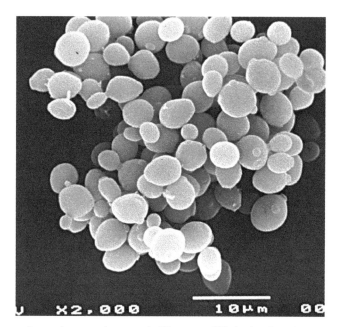

Fig. 3. Yeast (as seen under an electron microscope). (Courtesy of Katherine Smart)

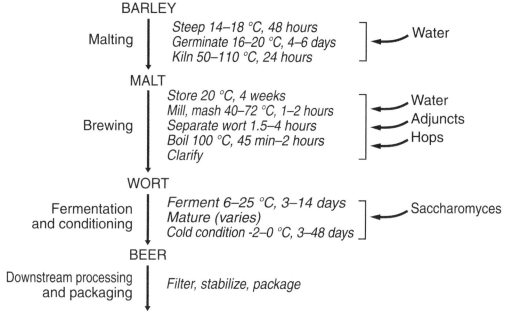

Fig. 4. Overview of malting and brewing.

fermentations has been mopped up by yeast. The yeast is harvested for use in the next fermentation. Beer is the only alcoholic beverage for which yeast is recycled in this way.

In *traditional ale brewing*, the beer is now mixed with hops, some priming sugars, and isinglass finings from the swim bladders of certain fish, which settle out the solids in the cask.

In *traditional lager brewing*, the "green beer" is matured by several weeks of cold storage prior to filtering.

Today, the majority of beers, both ales and lagers, receive a relatively short *conditioning* period after fermentation and before filtration. This conditioning is ideally performed at –1 to –2°C for a minimum of 3 days, under which conditions more proteins drop out of solution, making the beer less likely to go cloudy in the package or glass. Various stabilization treatments may be used.

The filtered beer is adjusted to the required carbonation before packaging into cans, kegs, or glass or plastic bottles.

The essentials of malting and brewing are summarized in Figure 4 and Table 1.

Table 1. The essentials of malting and brewing

Process Stage	Description
Selection of barley	Malting barleys (moisture content <12%) of relatively low total protein content (e.g., less than 10.5%), with high viability, released from dormancy and with endosperm with mealy texture that hydrates readily and possesses cell walls that are readily degraded
Malting	
Steeping	Staged addition of water at 14–18°C separated by air rests to raise moisture content to 43–46%
Germination	Controlled germination for 3–6 days at 16–20°C to degrade the endosperm cell walls and much of the protein
Kilning	Drying of malt at successively increasing temperatures (maximum for mainstream malts 105°C) to dry the grain (target <6% H_2O), while retaining much of the enzymic activity and developing color and flavor
Malt storage	For 2–4 weeks to avoid processing problems in the brewery, notably reduced rates of sweet wort collection.
Milling and mashing	Generation of particles accessible to mashing water. Mashing is often started at about 50°C (for 20 min) to allow remaining action of thermo labile β-glucanase and then raised to 65°C (e.g., for 1 hr) for starch gelatinization and action of amylase complex. Wort is then separated from grains.
Boiling	Wort is boiled with hops or hop preparations, typically for 1 hr. Clarification stage follows to remove "hot break" and residual hop material. Wort is cooled and air or oxygen added.
Fermentation	Ales typically fermented at higher temperatures (15–25°C) and therefore faster than lagers (6–15°C). Time range 3–14 days. Fermentation at a targeted rate of specific gravity drop and to a target "attenuation." Also diacetyl and pentanedione, which afford butterscotch/honey aromas, must be removed by prolonged contact of yeast with "green beer."
Maturation, stabilization, and packaging	Minimum regime is –1°C, usually for 2–3 days. Some hold longer. Filtration (kieselguhr or perlite-based). Removal of haze precursors by polyvinylpolypyrrolidone (polyphenols) and or silica hydrogels or tannic acid or papain (proteins). Removal of any contamination by pasteurization or filtration. Adjustment of CO_2 content and then filling of vessels

Further Reading

Briggs, D. E., Boulton, C. A., Brookes, P. A., and Stevens, R. (2004) Brewing Science and Practice. CRC Press, Boca Raton, FL, and Woodhead Publishing, Cambridge, UK.

Kunze, W. (2004) Technology: Brewing and Malting. Versuchs- und Lehranstalt für Brauerei, Berlin, Germany.

3. Beer Types

Beers can be classified in various ways:
1. overall style: ales or lagers
2. strength, in terms of "original extract" or alcohol content
3. color
4. principal grist ingredient, e.g., wheat beer, sorghum beer
5. region of (original) production, e.g., Pilsen, Burton ale, Irish stout
6. technological influence, e.g., dry beer, light beer, ice beer, flavor ingredient

Style

Traditionally, beers have been categorized into ales and lagers (Table 2).

Ales were traditionally fermented from relatively dark grists (featuring well-modified, quite highly kilned malts) using yeasts that rose to the top of the fermenting vessel (top fermenters). They were served at relatively warm temperatures (10–20°C) and contained comparatively low levels of carbon dioxide.

Lagers were traditionally fermented at lower temperatures from lightly colored grists (relatively undermodified, gently kilned malts) by yeasts that settled to the base of fermenting vessels (bottom fermenters). They were stored for protracted periods in the brewery before sale, during which time high carbon dioxide contents were developed, and they were dispensed at lower temperatures (0–10°C).

There is an increasing blurring of the boundaries between ales and lagers. Today, ales are often produced in deep cylindroconical vessels in which the yeast sinks to the bottom of the fermenters. Lager strains may be used to produce ale beers and vice versa. The same pale malt may be used for both types of beer.

Strength

It is customary in brewing to describe beers in terms of their original extract (OE; in some countries this is called original gravity, OG). This is a measure of the specific gravity of the worts at the start of fermentation. Thus, a beer may be said to be 10°Plato, for instance.

For most beers, in which fermentation is allowed to proceed to its natural conclusion, there is a direct proportionality between the OE and alcoholic strength. However, it is quite possible to produce worts of high strength but whose carbohydrates are not fermentable by yeast. In this case, a low-alcohol beer is produced at the end of fermentation. Likewise, the carbohydrates in some beers are rendered fully

Table 2. Major beer styles[a]

Style	Origin	Notes
Ales and stouts		
Bitter (pale) ale	England	Dry hop, bitter, estery, malty, low carbonation (on draught), copper color
India Pale Ale	England	Similar to bitter ale, but substantially more bitter
Alt[b]	Germany	Estery, bitter, copper color
Mild (brown) ale	England	Darker than pale ale, malty, slightly sweeter, lower in alcohol
Porter	England	Dark brown/black, less "roast" character than stout, malty
Stout	Ireland	Black, roast, coffee-like, bitter
Sweet stout	England	Caramel-like, brown, full-bodied
Imperial Stout	England	Brown/black, malty, alcoholic
Barley wine	England	Tawny/brown, malty, alcoholic, warming
Kölsch	Germany	Straw/golden color, caramel-like, medium bitterness, low hop aroma
Weizenbier[c]	Germany	Hefeweissens retain yeast (i.e., turbid). Kristalweissens are filtered. Very fruity, clove-like, high carbonation
Lambic	Belgium	Estery, sour, "wet horse-blanket," turbid. Lambic may be mixed with cherry (kriek), peach (peche), raspberry (framboise), etc. Old lambic blended with freshly fermenting lambic is called *gueuze*
Saison	Belgium	Golden, fruity, phenolic, mildly hoppy
Lagers		
Pilsner	Czech Republic	Golden/amber, malty, late hop aroma
Bock	Germany	Golden/brown, malty, moderately bitter
Helles	Germany	Straw/golden, low bitterness, malty, sulfury
Märzen[d]	Germany	Diverse colors, sweet malt flavor, crisp bitterness
Vienna	Austro-Hungary	Red-brown, malty, toasty, crisply bitter
Dunkel	Germany	Brown, malty, roast-chocolate
Schwarzbier	Germany	Brown/black, roast malt, bitter
Rauchbier	Germany	Smokey
Malt liquor	United States	Pale color, alcoholic, slightly sweet, low bitterness

[a] From Bamforth, C. (2005) Food, Fermentation and Micro-organisms, Blackwell, Oxford, UK.
[b] Meaning "old."
[c] Wheat beer.
[d] Meaning "March," for when it is traditionally brewed.

fermentable, in which case more alcohol is produced per unit of OE than in conventional beers.

For these reasons, it is increasingly common to describe beer strength by alcohol content, a value that gives a direct guide to the potency of a beer.

The term *low-alcohol beer* means different things in different countries, but one useful definition is that a product containing less than 2% ABV is low alcohol and one containing less than 0.1% is non-alcoholic.

Color

Ales are traditionally divided into pale ales (called "bitter" when on draft dispense), brown ales ("mild" on draft), and porters and stouts. These styles are progressively darker.

Most lagers are pale and are broadly classified into categories such as Pilsner, Helles, and Marzen, but some are dark, e.g., Dunkel and Schwarzbier.

Grist

The majority of beers in the world are brewed from barley malt (perhaps supplemented with some adjuncts). However, some beers are made from malted wheat and are called *Weizenbier* or *Weissbier*. In Africa, much of the beer is made from sorghum.

Region

Terms such as "Pils" or "Pilsner" have become generally synonymous with mid-strength lagers and no longer have a strict definition. Time was when they would have been restricted to a style of lager produced in the Pilsen district of the old Bohemia. The same situation applies to Burton ale. However, many beers are marketed on their regional provenance, e.g., San Francisco Steam Beer and Newcastle Ale.

Technology

Perhaps the most significant classification of beers according to a technological application are the light (lite) beers, in which steps are taken to maximize the fermentability of the wort and thereby obtain a low-carbohydrate product.

Dry beers are a related concept.

The term "draft beer" means different things in different markets. Generally, it refers to beer dispensed from kegs through a tap, but it has come to be used for canned or bottled beer that has not been through a pasteurization process.

Ice beer has been subjected to a process wherein tiny ice crystals are produced and then either removed or melted back into the beer downstream during the process.

The most intense research and development activity in many brewing companies centers on new product development. From such activity in recent years we have seen black lagers, colorless lagers, "nitrokegs," and all manner of flavored beers (e.g., oyster stout, heather ale, citrus lagers, etc.). The "malternatives" are produced from lightly colored grists of malt and sugars fermented prior to the addition of diverse flavors.

Further Reading
Classic Beer Styles series, Brewers Association, Boulder, CO.

4. The Quality and Wholesomeness of Beer

The quality of beer can be divided into several categories:
1. flavor
2. foam
3. color
4. clarity
5. wholesomeness

The characteristics of the package (e.g., the labeling, bottle shape, etc.) are not discussed here, but they have a profound effect on the consumer's perception of the product.

Flavor

The substances in beer that contribute to flavor can originate from the malt and adjuncts, the hops, or the water or are products of yeast metabolism during fermentation. If the process stream or the product is infected, flavor-active materials may also originate from spoilage bacteria or yeasts. The flavor of beer is summarized in the flavor wheel of the American Society of Brewing Chemists (Fig. 5).

The tongue detects basically four flavors: sweet, sour, salty, and bitter (Fig. 6). All other flavors are detected by smell (Fig. 7). The sensation caused by carbonation, which is essentially pain recognition (pleasurable!), is detected by the trigeminal nerve. This, alongside astringency, is a part of the mouthfeel character of beer that remains one of the more challenging areas for scientific investigation of beer quality.

The Japanese speak of *umami* (a character of the type imparted to Chinese food by monosodium glutamate), but there have been no comprehensive studies of this flavor in relation to beer. Brewers have from time to time searched for a positive drinkability factor but have succeeded only in identifying characteristics that suppress drinkability (e.g., excessive hop character and diacetyl).

Sweetness

Sweetness in beer is derived from sugars, either those from malt and other grist materials and those that survive fermentation or those that are added by the brewer ("primings") to balance sweetness and bitterness.

Sourness

Sourness is caused by the H^+ ion and is therefore imparted by acids, such as citric and acetic acid that can come from the malt but are also products of yeast metabolism, or those such as lactic or phosphoric acid that may be deliberately introduced by the brewer.

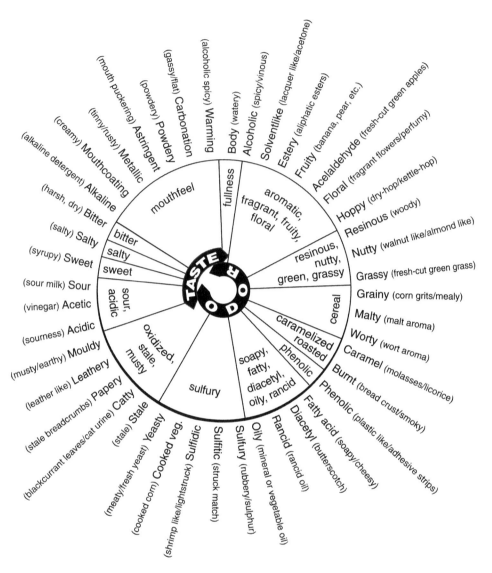

Fig. 5. Beer flavor wheel. (Redrawn from Meilgaard, M. C., Dalgliesh, C. E., and Clapperton, J. F. 1979. Beer flavor terminology. J. Am. Soc. Brew Chem. 37:47-52.)

Beer is a relatively well-buffered material, and most beers have a pH of 4.0–4.6. Those with a lower pH, e.g., the lambic/gueuze products of Belgium, have potentially greater sourness.

Salt

Salts in beer originate from the grist and from the water. Ions, such as bicarbonate, have a direct effect on pH. Sodium directly impacts the "saltiness" of a beer, as does potassium. Sulfate allegedly causes dryness and chloride palate fullness, but the precise relevance of the chloride:sulfate ratio (which often features in beer specifications) has not been proven. Iron is directly detectable as a metallic flavor.

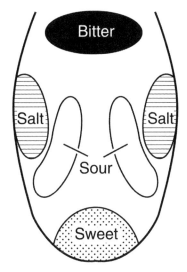

Fig. 6. Concentration of taste receptors on the tongue.

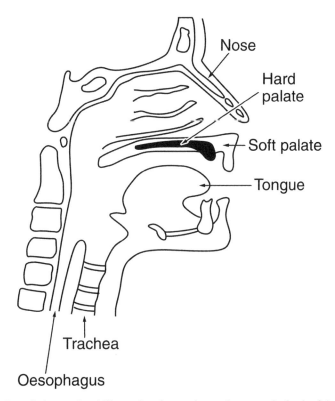

Fig. 7. Cross-section of a human head illustrating the continuum between the back of the throat and the nose.

Bitterness

The bitterness in beer is caused by iso-α-acids derived from the resin component of hops.

Aroma

A wide range of substances directly impact the aroma of beer.

From malt come materials produced during kilning, substances that afford roasted, burnt characters to a beer, as well as the flavor characteristically defined as "malty." Ales tend to possess higher levels of these notes than do lagers, because they are kilned to higher temperatures.

Dimethyl sulfide (DMS) derives from a heat-sensitive precursor in malt. This precursor is mostly lost during the kilning of ale malts but not during the kilning of lager malts. Therefore, DMS tends to survive into many lagers.

From hops come the essential oils. These are very volatile and tend to blow away during the wort boiling operation. In traditional European lager brewing, a proportion of the hops is held back until nearly the end of boiling to allow some of the oils to survive and give a characteristic aroma to the beer: the so-called "late hop aroma." In traditional English ale brewing, whole hops are added to the finished beer as a source of a strong, dry hop character.

Yeast produces a wide range of flavor-active materials. These include sulfur-containing substances such as hydrogen sulfide (made from sulfate and from sulfur-containing amino acids), DMS, mercaptans, and many others, especially produced by lager strains of yeast at low temperatures. The sulfur compounds give flavors described by terms such as cooked vegetable, canned corn, and rubbery.

Yeast produces alcohols in addition to ethanol, such as iso-amyl alcohol and 2-phenylethanol. The yeast esterifies these to yield still more flavor-active esters, such as iso-amyl acetate (banana) and phenylethyl acetate (honey).

During fermentation, yeast naturally produces a substance called diacetyl (often referred to as vicinal diketone, VDK) that has an intense butterscotch character. Yeast subsequently removes the diacetyl—and this is why in conventional fermentation practice it is necessary to leave yeast in contact with the "green" beer to "mop up" the diacetyl.

By introducing "tingle" to beer, CO_2 impacts the mouthfeel of beer. Other factors that influence the texture of beer are not well understood, although it is known that nitrogen gas, added in some parts of the world (especially in the United Kingdom and Ireland) to enhance foam, causes a smoothing of the palate and a suppression of hop aroma. This illustrates that there are interacting factors that influence the flavor of beer.

One character may suppress another (e.g., phenylethanol suppresses the perception of DMS).

The flavor of beer is also influenced by visual stimuli. For example, the same beer scores differently for flavor if it is served with a head or without a head, and the addition of a flavorless coloring agent to make a lager look like an ale leads people to "score" it as an ale.

The flavor of beer is not static. It changes in the package, generally for the worse. In particular, a cardboard character will develop. The most significant factors influencing this deterioration in flavor are temperature (it is best to store beer cold) and the presence of oxygen, which is why brewers pay especial attention to minimizing oxygen levels in the container.

Foam

The amount of foam that can be produced on a beer is in direct proportion to the amount of CO_2 in it. This applies equally to a glass of soda—yet the foam on the soda does not last. Foam lasts on beer because it contains surface-active materials that move into the bubble walls and form a framework to stabilize them.

The principal backbone material for these bubbles is protein, which originates in the malt. Those proteins that are relatively hydrophobic (water-hating) prefer to leave the beer and enter the foam. There they interact with another set of hydrophobic materials, the iso-α-acids (bitter substances derived from hops), to produce stable foams.

Other surface-active materials tend to destabilize foam by interfering with these protein–iso-α-acid interactions. These include lipids and detergents (particularly those introduced into the glass during washing or dispense) and ethanol (the higher the alcohol content of a beer, the more unstable the foam). However, the first 1% of alcohol in a beer is important as it promotes foaming by reducing surface tension.

Actually, the level of CO_2 in most beers is higher than its direct solubility; thus, the beer is said to be super-saturated. However, foaming is usually not spontaneous. It occurs only if a process called *nucleation* occurs through agitation during the act of pouring, use of scratched glasses, etc. In some beers, unwanted nucleation sites are derived from raw materials (e.g., intensely hydrophobic peptides from fungal contaminants of grain, oxidation products of hops, or traces of heavy metals), and foaming occurs immediately when a can or bottle is opened. This is called *gushing*.

Color

Most of the color in beer results from melanoidins, which are complex molecules formed by the reaction of sugars and amino acids during the Maillard reaction. This occurs during the heating stages, especially the kilning of malt and, to a lesser extent, the boiling of wort. Thus, lager malts used for lightly colored beers receive less intense kilning than malts used in the production of ales.

Some color formation may also be caused by the oxidation of polyphenols in the brew house (a reaction analogous to the browning of sliced apples).

Color is sometimes adjusted downstream by the addition of caramels, but increasingly common for this purpose is the use of extracts from roast malts.

Clarity

The polymerization of polyphenols caused by oxidation not only results in color formation, but it also contributes to haze production through the interaction of the polymerized polyphenols with proteins. There are other sources of turbidity in beer, including polysaccharides, oxalate, and dead bacteria from the grist; yeast leachates; foam stabilizers; etc.

Wholesomeness

In beer there may be
1. components of beer that are potentially beneficial to health
2. components of beer that are potentially detrimental to health

Beneficial Components

Ethanol itself is now recognized as being beneficial, provided it is not taken to excess. At least part of the effect must be through a calming influence on the body.

Several studies have revealed a so-called "U-shaped" or "J-shaped" curve relationship between the mortality risk and consumption of alcohol, in which the death rate declines as alcohol consumption increases to approximately 26 units of beer per week (approximately two pints of beer per day) for men and 80% of this rate for women (Fig. 8). At higher rates the risk of death increases. Such relationships apply to all forms of alcoholic drink.

In particular, the risk of death from cardiovascular disease declines as alcohol consumption increases. It is believed that alcohol reduces the clogging of arteries by lipids and also reduces the tendency of blood clots to form.

Beer is, of course, largely water and to an extent counteracts the dehydration caused by alcohol, which is particularly prevalent for high-alcoholic drinks. Low-alcohol beers are alternatives to formulated isotonic drinks.

The sugar content of most beers is much lower than that of most non-diet soft drinks, and light beers are proportionately lower in calories.

Beer is essentially fat-free.

Beer is a useful source of the B vitamins niacin, pyridoxine, riboflavin, and B_{12} and may constitute a particularly valuable source of folate. It is also a respectable source of antioxidant materials.

Beer is a good source of minerals such as magnesium and potassium and has a low sodium:potassium ratio, which is beneficial for blood pressure.

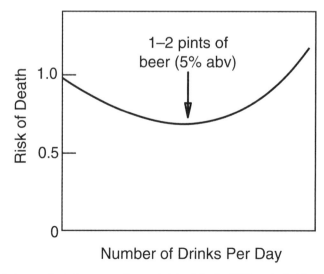

Fig. 8. Relationship between alcohol consumption and risk of death. ABV = alcohol by volume.

Beer is a relatively inhospitable environment for the growth of microorganisms and in particular does not support the growth of pathogens.

Potentially Adverse Components

Excessive consumption of alcohol should be avoided. Not only can it induce antisocial behavior, but it has been associated with increasing the risk of death through problems such as increased blood pressure, cirrhosis of the liver, and perhaps cancers.

In common with all other foodstuffs, materials have been identified in beers that are questionable in terms of health. The incidence of these occurring increases as analytical techniques become more and more sensitive. In all cases where food safety issues of this type have been identified, the malting and brewing industries have responded by establishing procedures to minimize levels of the components concerned.

Materials in this category include

1. Pesticides. It is frequently necessary in the cultivation of barley and hops to preclude development of undesirable pests, such as fungi. The pesticides used are always screened for their implications for health and process. Residue levels are invariably low because they are washed off during steeping of barley.
2. Nitrosamines and apparent total nitroso compounds (ATNCs) (discussed in Chapter 5).
3. Monochloropropanols may be produced during the intense heating of any material that contains glycerol and chloride (Fig. 9). Therefore, they are found in very dark malts, but they do not survive into beer.
4. Mycotoxins, such as deoxynivalenol (vomitoxin) from *Fusarium* spp., are poisonous substances produced by fungi, especially molds (Fig. 10). Ergo, there is a benefit to the use of carefully regulated and screened fungicides.
5. Lead and other heavy metals. There is a need for attention to pipe construction and composition of ingredients.

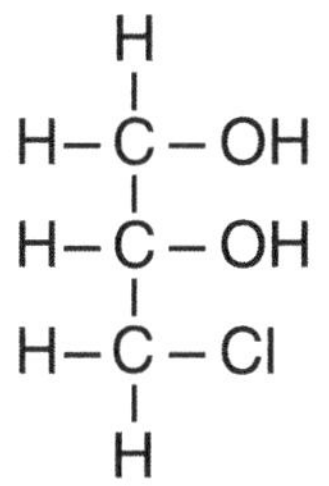

Fig. 9. Chemical structure of monochloropropanol.

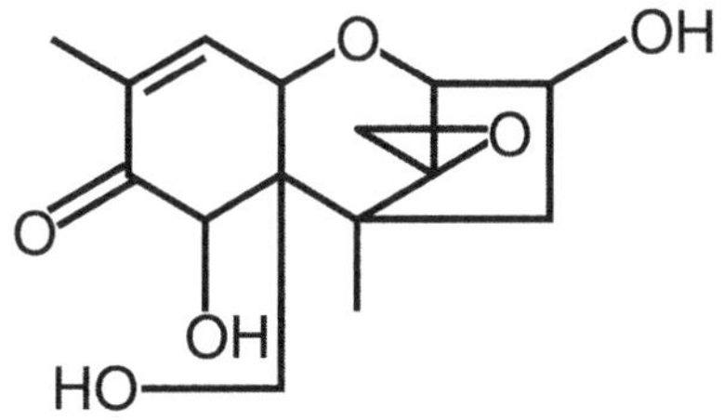

Fig. 10. Chemical structure of deoxynivalenol (vomitoxin).

6. Nitrates. Nitrates are often found in fertilizers and are present at higher levels in water and other raw materials from agricultural regions.

Further Reading

Bamforth, C. W. (2004) Beer: Health and Nutrition. Blackwell Publishing, Oxford, UK.

Baxter, E. D., and Hughes, P. S. (2001) Beer: Quality, Safety and Nutritional Aspects. Royal Society of Chemistry, Cambridge, UK.

5. Barley and Malting

Barley

Although it is possible to make beer using raw barley and added enzymes (so-called "barley brewing"), it is extremely unusual in this day and age.

Unmalted barley alone is unsuitable for brewing beer because

- it is hard and difficult to mill
- it lacks most of the enzymes needed to produce fermentable components in wort
- it contains complex viscous materials that slow down solid–liquid separation processes in the brewery and that may cause clarity problems in beer
- it contains unpleasant raw and grainy characters and is devoid of pleasing flavors associated with malt

Barley belongs to the grass family (Fig. 11). Its Latin name is *Hordeum vulgare*. Some people retain the use of this name for six-row barley, using the term *Hordeum distichon* for two-row barley.

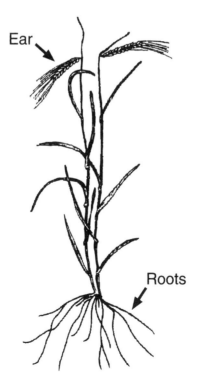

Fig. 11. Barley plant.

22 / Chapter 5

Fig. 12. Two-row (left) and six-row (right) barley. (Courtesy of NIAB, Cambridge, U.K.)

The part of the plant of interest to the brewer is the grain on the ear (Fig. 12). Sometimes this is referred to as the seed, but individual grains are generally called kernels or corns.

Four components of the kernel (Fig. 13) are particularly significant:
1. the embryo, which is the baby plant
2. the starchy endosperm, which is the food reserve for the embryo
3. the aleurone layer (two to four cells thick), which generates the enzymes that degrade the starchy endosperm
4. the husk (hull), which is the protective layer around the corn.

Barley is unusual among the cereals in that it retains the husk after threshing. This tissue is traditionally important for its role as a filter medium in the brew house when the wort is separated from spent grains.

When a barley corn takes up moisture, the embryo and endosperm are hydrated. Hydration switches on the metabolism of the embryo, which sends hormonal signals to the aleurone layer, triggering the synthesis of enzymes responsible for digesting the components of the starchy endosperm. The digestion products migrate to the embryo and sustain its growth.

The aim is controlled germination to soften the grain, remove troublesome materials, and expose starch without promoting excessive growth of the embryo, which would be wasteful (*malting loss*). The three stages of commercial malting are
1. *steeping*, which brings the moisture content of the grain to a level sufficient to allow metabolism to be triggered in the grain

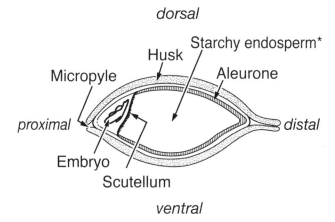

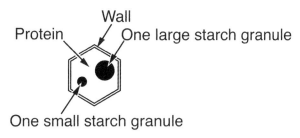

Fig. 13. Top, cross-section of a barley kernel and bottom, a simplified representation of a single cell from the starchy endosperm. Note: only one large and one small starch granule are depicted. There are thousands of each.

2. *germination*, during which the contents of the starchy endosperm are substantially degraded (called *modification*), resulting in softening of the grain
3. *kilning*, during which the moisture is reduced to a level low enough to arrest modification

The embryo and aleurone are both living tissues, but the *starchy endosperm* is dead. The latter is a mass of cells, each comprising a relatively thin cell wall (approximately 2 μm thick) inside which are packed many starch granules within a matrix of protein. These are the food reserves for the embryo. However, the brewer's interest in them is as the source of fermentable sugars and assimilable amino acids that the yeast will use during alcoholic fermentation.

The wall around each cell of the starchy endosperm comprises 75% β-glucan, 20% pentosan, 5% protein, and some acids, including ferulic acid and acetic acid.

The β-glucan comprises long, linear chains of glucose units joined through β-linkages (Fig. 14). Approximately 70% of these linkages are between C-1 and C-4 of adjacent glucosyl units (so-called β 1-4 links, just as in cellulose), and the remainder are between C-1 and C-3 of adjacent glucoses (β 1-3 links, which are NOT found in cellulose). These 1-3 links disrupt the structure of the β-glucan molecule and make it less ordered, more soluble, and more digestible than cellulose.

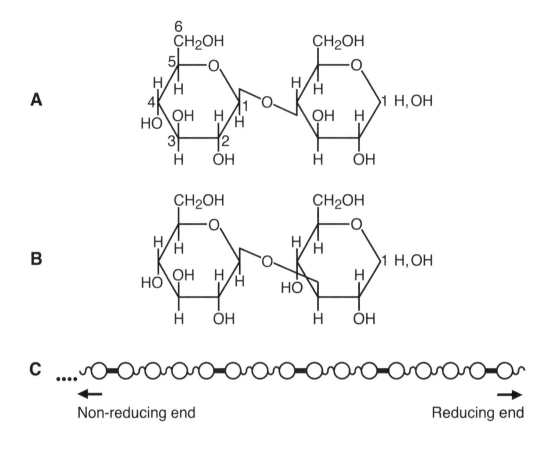

Fig. 14. Structural elements of β-glucan in barley. A, Cellobiose, the β 1-4 repeating unit; B, laminaribiose, the β 1-3 repeating unit; C, schematic representation of barley β-glucan structure, ~β 1-4; –β 1-3; and D, a cellulosic region. For an explanation of reducing and non-reducing ends, see Appendix 1.

However, there are regions in the barley β-glucan molecule that comprise 10 or more adjacent β 1→4 links. These "cellulosic" regions contribute to the insolubility of the glucan, which may also be due to associations between the glucan and other components of the cell wall.

The proportion of the β-glucan in the cell wall that is relatively insoluble is called "hemicellulose"; that which is relatively easily extracted by warm water is referred to as "gum." Solutions of both gum and solubilized hemicellulose are very viscous and therefore need to be degraded if problems with solid–liquid separation are to be avoided in the brewery. The β-glucan also needs to be degraded if the starch and the protein are to be exposed.

Much less is known about the *pentosan* (arabinoxylan, Fig. 15) component of the cell wall, and it has long been believed that it is less easily solubilized and broken down than the β-glucan and that substantial amounts of it largely remain in the spent grains after

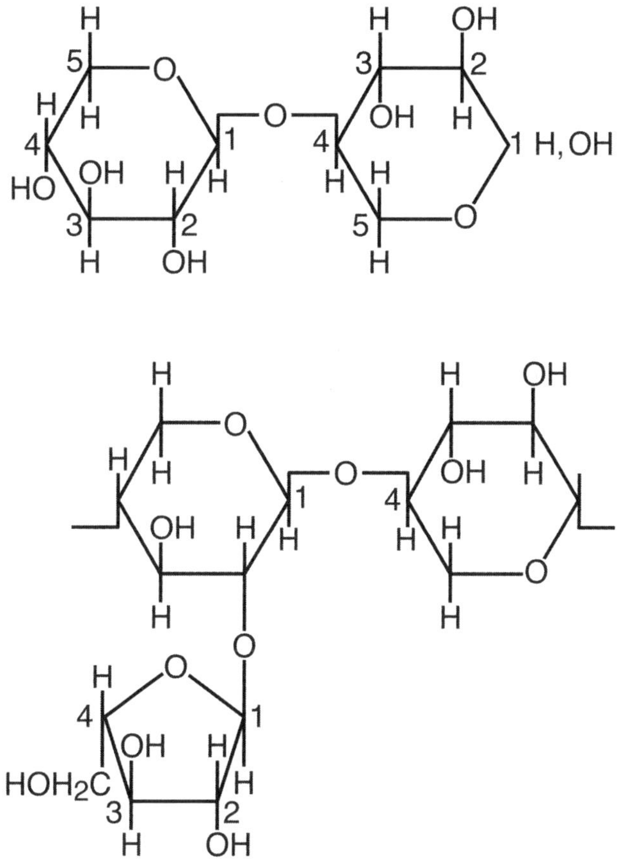

Fig. 15. Key linkages in arabinoxylans in barley endosperm. Top, β 1-4-linked xylosyls; bottom, arabinose-linked α 1-2 to xylosyl.

mashing. Recent work has refined this view (see Part 6). Arabinoxylan comprises a backbone of multiple xylosyl residues, with occasional side chains of individual arabinosyl residues. Ferulic acid and acetic acid are also covalently attached.

The *starch* in the cells of the starchy endosperm is in two forms: large (A) granules (approximately 25 μm in diameter) and small (B) granules (5 μm). These granules may be associated with both protein and lipid, both of which may limit the degradability of the starch.

The proteins in the starchy endosperm may be classified according to their solubility characteristics. The two most relevant classes are the *albumins* (water-soluble) and the *hordeins* (alcohol-soluble). In the starchy endosperm of barley, the hordeins are quantitatively the most significant. They are the storage proteins, and they need to be substantially degraded in order that the starch can be accessed.

Selection of Barley

It is possible to make malt from any barley. However, in practice, the brewer, and therefore the maltster, is very choosy about the barley that is used. Fundamentally, it

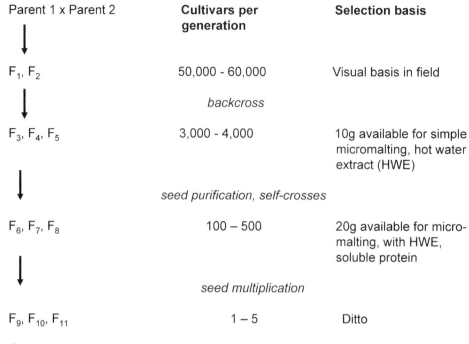

Fig. 16. Simplified barley-breeding program. (Redrawn from Bamforth, C. W., and Proudlove, M. O. 1992, The evaluation of barley for malting. Pages 90-106 in: Home-Grown Cereals Authority Conference on Cereals R & D, HGCA, London.)

should be free from foreign matter, infection, infestation, odor, and taint, but the following parameters are also important.

Variety. Top-quality malts are produced from those varieties that modify rapidly and evenly and yield high levels of fermentable extract. These are the so-called "malting varieties" as opposed to "feed barleys."

In Europe, most barleys are the *two-row* variety. Only one kernel develops at each node on the ear, and it appears as if there is one kernel on either side of the axis of the ear.

In North America, both two-row and *six-row* varieties are grown. The six-row varieties develop three corns per node. Obviously, there is less room for the individual kernels, and they tend to be somewhat twisted and smaller. Six-row varieties tend to have higher N contents and therefore higher potential for developing enzymes. However, their pro rata yield of extract must be lower and the proportion of husk is greater.

In many countries, there are programs through which new and improved barley varieties are developed, tested, and recommended (Fig. 16). In the United States, this is through the American Malting Barley Association (AMBA). In such programs, the aims include ever-increased yields of extractable material ("extract"), varieties that cause fewer processing problems, increased resistance to infection, and shorter straws to prevent the plant from falling over ("lodging") in the field (Table 3).

Table 3. Guidelines to breeders of malting barley[a]

	Two-Row Barley	**Six-Row Barley**
Barley factors		
Plump kernels (retained on 6/64-in.screen)	>90%	>80%
Thin kernels (through 5/64-in. screen)	<3%	<3%
Germination (4 ml, 72 hr, Germinative Energy)	>98%	>98%
Protein	11.5–13.0%	11.5–13.5%
Skinned and broken kernels	<5%	<5%
Malt factors		
Total protein	11.3–12.8%	11.3–13.3%
Retained on 7/64-in. screen	>70%	>60%
Measures of Malt modification		
Beta-Glucan (ppm)	<115	<140
Fine-Coarse Difference	<1.5	<1.5
Soluble/total protein	42–47%	42–47%
Turbidity (NTU)	<10	<10
Viscosity (absolute centipoise)	<1.50	<1.50
Congress wort		
Soluble protein	4.9–5.6%	5.2–5.7%
Extract (fine grind, dry basis)	>81.0%	>79.0%
Color (°ASBC)	1.6–2.0	1.8–2.2
Malt enzymes		
Diastatic power (°ASBC)	120–160	140–180
Alpha amylase (Dextrinizing Units)	45–80	45–80

General comments

  Barley should mature rapidly, break dormancy quickly without pregermination, and germinate uniformly.
  The hull should be thin and bright and adhere tightly during harvesting, cleaning, and malting.
  Malted barley should exhibit a well-balanced modification in a conventional malting schedule with 4-day germination.
  Malted barley must provide desired beer flavor.

[a] Reproduced with permission from the American Malting Barley Association (http://www.ambainc.org/ni/Guidelines%202004.pdf).

Experts can look at barley kernels and identify the varieties. This is clearly valuable if there is a suspicion that a given batch of barley comprises a mixture. Factors that an expert is looking for include the extent of wrinkling of the husk, the length of the rachilla (an appendage emerging from the embryo end of the grain within the ventral furrow; not depicted in Figure 13) and whether or not it is hairy, and the color of the aleurone (white or blue) in pearled grain (to "pearl" is to remove the outer layers).

Alternatively, the protein or DNA of grain can be extracted and separated by polyacrylamide gel electrophoresis; the resultant pattern of bands is characteristic of the variety.

Nitrogen content. The brewer wants a high yield of sugar from malt, and this is derived from starch. For a kernel of a given size, there is more room for starch if the amount of protein is relatively low. Maltsters and brewers frequently describe the amount of protein in terms of nitrogen, with N × 6.25 indicating the relationship between N and protein. (Recent studies with some cereals suggest that the equation should be N × 5.7.)

Maltsters pay a "malting premium" for the right barley variety grown to have the desired level of protein. To achieve a low enough N content, the farmer is restricted in how much fertilizer can be used. Therefore, the yield of malting barleys is less than that of feed barleys and hence the price is higher.

There must be some protein present, as this is the fraction of the grain that includes the enzymes and is the origin of amino acids (for yeast metabolism) and foam polypeptide. The amount of protein needed in malt depends on whether the brewer intends to use some adjunct material as a substitute for malt. For example, corn syrup is a rich source of sugar but not of amino acids, which need to come from the malt.

Typical N levels in a two-row European variety may be 1.5–1.7%, but for a North American six-row variety may more typically be 1.8–2.0%. This means that six-row barleys develop high enzyme levels (enzymes are proteins) and malt from six-row varieties is more likely to be used when converting starchy adjuncts.

Moisture. Harvested barley contains water, the level of which depends on the growth location. Generally in North America, the harvest is not especially wet, but in northern Europe there is usually a need to dry barley immediately after harvest to prevent the growth of fungi and to prevent it from pregerminating. A specification for moisture content is likely to be <12%. The maltster does not want to buy moisture, and high moisture content results in reduced stability.

Kernel size. Thinner corns take up water more rapidly but also contain less starch and proportionally more husk. For a two-row barley, a specification may be that 60% of kernels should be retained by a 2.8-mm sieve, >85% retained on a 2.5-mm sieve, and >97% on a 2.0-mm sieve. (Any material passing through the last of these is referred to as "screenings.")

Another useful index of size is the thousand kernel weight (TKW), which is literally the mass of a thousand kernels; higher values signify fatter corns.

Modification potential. Dead grain will not germinate. Therefore, a critical parameter is the viability (or germinative capacity) of a batch of grain. This is assessed by measuring the numbers of grains that germinate ("chit") over 3 days in the presence of hydrogen peroxide. Staining the embryo with a tetrazolium dye (2,3,5-triphenyl tetrazolium chloride) makes for a more rapid evaluation: living grain is stained by this material, but dead kernels are not. This test can be carried out on deliveries of grain as they arrive at the malt house, as can determination of moisture and N content by techniques such as near infrared spectroscopy. A typical specification for germinative capacity is >96% (i.e., no more than four grains in every 100 should be dead).

To achieve consistent and evenly modified malt, however, it is desirable that all the grain in the batch modifies to the same extent. Even though a kernel of barley may be viable, it may differ from other kernels in its *vigor*. In the extreme, a barley sample can be *dormant*, meaning it will not germinate even though it is alive. The mechanism that controls dormancy is unknown, but it can be readily evaluated by counting the numbers of corns that start to germinate after 3 days in petri dishes containing filter paper and 4 ml of water. This is the germinative energy test.

Dormancy is an important phenomenon because it is the mechanism by which barley prevents the grain from germinating on the ear. *Pregermination* does sometimes occur, e.g., during wet harvests, and it is a major problem because such grain is prone to infection and also is likely to be killed when the barley is dried after harvesting. Pregermination can be detected by staining grain with fluorescein dibutyrate. If the grain has started to germinate, it has made the enzyme that breaks down this compound, and this can be detected through the release of a fluorescent material.

A maltster will not attempt to malt barley unless it displays a germinative energy in excess of 96%. Usually, grain emerges from dormancy quite rapidly (in less than 3 months), but some varieties from cool climates are especially prone to dormancy, which can be broken by storage under warm (30°C) conditions.

The vigor of nondormant grain can be evaluated in the germinative energy test by counting the numbers of corns that chit after 1, 2, and 3 days. As grain is stored, it tends to chit more readily, and this increased vigor requires that the maltster alter the malting conditions if a malt of the necessary specification is to be achieved.

Water sensitivity is induced when barley is exposed to too much water and can be evaluated by carrying out the germinative energy test with 8 ml of water. Water-sensitivity is probably the result of a "swamping" of the grain, which prevents ready access to the oxygen the embryo needs to support respiration. For this reason, barley is steeped in "interrupted" programs, in which relatively short steeps are interspersed with "air rests." There is a belief that water sensitivity arises because of competition for oxygen between the grain and its microflora.

Barley kernels can be divided into those with *mealy* endosperms, which have a loose crumbly texture that allows easy water uptake and digestion, and those with *steely* endosperms, which are vitreous and hard and which take up water with greater difficulty (Fig. 17). The β-glucans and proteins are present in larger quantities, are more closely packed and, as a consequence, are more intractable in steely endosperm.

An expert can slice longitudinally through a barley corn and visually inspect it to differentiate between these types of endosperms. Alternatively, light shone through dehulled grain reveals via transflectance that the mealy endosperms appear darker than the steely ones. Better varieties tend to have mealy endosperms. In practice, many individual grains will have both mealy and steely regions within their endosperms.

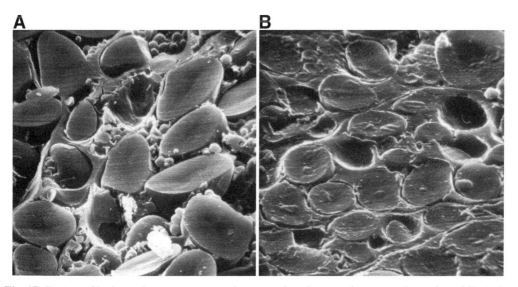

Fig. 17. Texture of barley endosperm as seen under a scanning electron microscope. A, mealy and B, steely. (Courtesy of Brewing Research International)

Sampling

Individual deliveries of barley to a malt house contain many kernels and will enter a storage silo that may hold upwards of 30,000 tons. Given a TKW of 40 g, such a silo will contain 750 billion individual kernels, and there are many reasons for corn-to-corn variation (heterogeneity):
- different barley varieties
- different batches of the same variety grown in different locations
- different batches of the same variety grown during different seasons
- different grains on a single ear of barley
- differences within the endosperm of a single corn
- differences that occur because of changes in a given batch of barley during storage
- different malting runs with the same barley
- different times at which barley is transferred between stages in a batch malting operation (e.g., time of loading of green malt onto the kiln)

Obviously, it is necessary to have sampling regimes that allow analysis of grain that is as representative as possible from any shipment or stored batch. Spear ("trier") samplers (Fig. 18) that are 6–12 feet long are used to draw samples from various positions across and through a bed. The samples are then mixed thoroughly prior to analysis.

Agronomics and Farming Practices

In North America, barley grows in approximately 100 days. When growing malting barley, the farmer must not only apply good standard practices (e.g., weed control, ground preparation, and clean implements) but must also recognize the needs of the brewer (and therefore maltster) in respect of variety, low N content, the avoidance of certain pesticides, etc.

There is, of course, a strong environmental–varietal interaction. Some environments are better suited to the growing of malting barley and some varieties can be matched to specific environments, rather like the viticulturalist lines up cultivars with locations. Most barley in the Northern Hemisphere is sown between January and April and is referred to as Spring barley. The earlier the sowing, the better the yield and the lower the N content because starch accumulates throughout the growing season. In locales with mild winters, some varieties (Winter barleys) are sown in September and October. There is a prejudice in some countries against winter varieties, the suggestion being that they give inferior beer. This is unfounded.

Fig. 18. A trier.

The best yields of grain are in locales where there is a cool, damp growing season allowing steady growth and then fine, dry weather at harvest to ripen and dry the grain. Grain grown during very hot, dry summers is thin and poorly filled and has a high N content.

The major malt houses in North America are located near the main barley-growing areas: western Minnesota, North Dakota, and eastern South Dakota in the north central United States and Alberta, Saskatchewan, and Manitoba in Canada.

Physiology and Practice of Malting

Commercial malting operations comprise four basic stages:
1. barley reception (intake, drying, and storage)
2. steeping
3. germination
4. kilning

A maltster is likely to be producing a range of malts (for ale, lager, specialty, distilling, food, etc.) and strives to minimize energy usage, labor costs, and effluent demands while maximizing yield. To produce quality malts, it is impossible to avoid some embryo development, but the more this "malting loss" occurs, the less the yield of the malt.

Reception

Grain arrives at the malt house by road or rail, and before it is unloaded, the barley is weighed and a sample is tested for the key parameters of viability, nitrogen content, and moisture. Expert evaluation will also provide a view on how clean the sample is in terms of weed content and whether the grain "smells sweet" (i.e., is not obviously infected). A few grains may be sliced in half lengthways and their endosperms assessed to determine whether they are mealy or steely. This involves use of a farinator (Fig. 19).

Once accepted, the barley normally is cleaned, to remove everything from dust and weeds to dead rodents, and screened, to remove small grain and dust, before passing into a silo, perhaps via a drying operation in areas with damp climates. Grain should be dry to counter infection and outgrowth (pregermination).

Drying is seldom performed at temperatures greater than 55°C, and the temperature of the grain does not get above 30°C because of the latent heat of evaporation (i.e., the heat energy consumed by water to enable it to evaporate; as the water "siphons" off the heat, the grain remains relatively cool). A continuous throughput of air is used, and drying may be continuous or in a batch operation.

Dry barley may be stored in various locations, ranging from steel- or concrete-framed sheds capable of holding up to 30,000 tons to steel bins holding no more than 750 tons. Whichever facility is used, it is essential that the store is protected from the elements and that it be ventilated, because barley, like other cereals, is susceptible to various infections, e.g., from *Fusarium* spp., storage fungi such as *Penicillium* and *Aspergillus* spp., mildew, and to pests such as aphids and weevils.

Pesticides, approved for use on the basis of health and safety legislation, have an important role. Like anything else accumulating on the surface of barley, they are washed off during the steeping operation and so do not get into the malt used for brewing. Barley

32 / Chapter 5

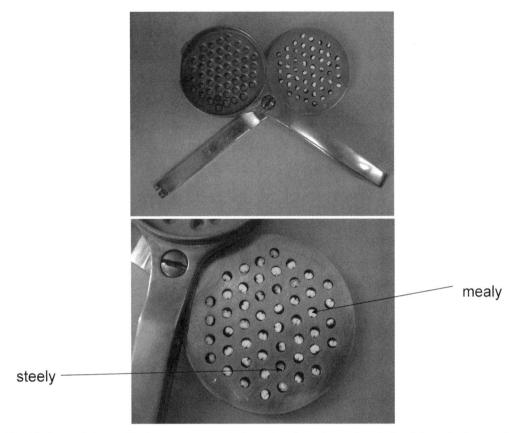

Fig. 19. Top, a farinator. The bottom photograph shows how it can be used to distinguish mealy from steely grains. (Courtesy of Stan Sole)

is crawling with organisms on, in, and below the husk. Most of these are benign and may actually play some beneficial roles, for example, by providing hydrolytic enzymes.

Steeping

Steeping is probably the most critical stage in malting. If homogeneous malt is to be obtained, then the aim must be to hydrate the kernels in a batch of barley evenly.

The steeping conditions required to achieve the desired moisture content differ depending on parameters such as variety, kernel size, protein content, and physiological status of the grain.

Steeping regimes are determined on a barley-by-barley basis by small-scale trials. Barleys differ in how much moisture they must contain in order to germinate. None germinate with moisture contents below 30%, and some (e.g., those with high N content or with steely endosperms) might need levels as high as 50%. For most barleys, the target is 42–46%. The grain swells in volume by about 50% during steeping.

The properties of the barley change over time. For example, it increases in vigor, which is reflected in its enhanced capability for synthesizing enzymes and, therefore, rate of modification of the endosperm. Thus, barley needs to be processed differently in the malt house depending on how long it has been in storage.

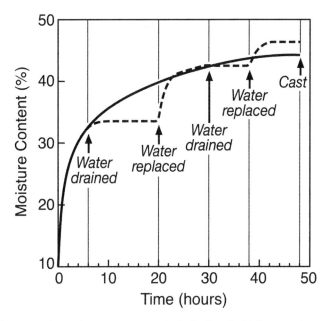

Fig. 20. Impact of interrupted steeping on water uptake in barley. Solid line, continuous steep; dotted line, interrupted steep.

Water cannot pass through the pericarp, the waxy layer of the grain that lies between the hull and the aleurone (not depicted in Fig. 13). In undamaged grain, water has only one substantial point of entry, the micropyle. The water entering through this region hydrates first the embryo and then the starchy endosperm.

Barleys that are easiest to malt take up moisture more readily and tend to have mealy rather than steely endosperms. Thinner kernels also take up water more rapidly, but contain more nitrogen and husk and less starch than do fatter kernels. In some locations, barley is graded before steeping according to size.

Uptake is more rapid as water temperature increases. The steep water is either drawn from a well, in which case it is likely to have a relatively constant temperature of 10–16°C, or it may come from a city supply, in which case a temperature-control system may have to be installed to heat or cool the water, depending on the season.

Barley also needs oxygen in order to support respiration in the embryo and aleurone. Because oxygen access is inhibited if the grain is submerged in water for excessive periods, interrupted steeping operations (Fig. 20) are used. Instead of submerging the barley in water and leaving it, the grain is steeped for a period of time and then removed for a so-called "air-rest" period. The barley is then submerged again, and the process is repeated. Air rests serve the additional purpose of removing carbon dioxide and ethanol, which suppress respiration.

A typical steeping regime may involve an initial steep to 32–38% moisture content (lower for more water-sensitive barleys). The start of germination is prompted by an air rest of 10–20 hr followed by a second steep to raise the water content to 40–42%. Emergence of the root tip (chitting) is encouraged by a second air rest of 10–15 hr before the final steep to the target moisture content. The entire steeping operation may take 48–52 hr.

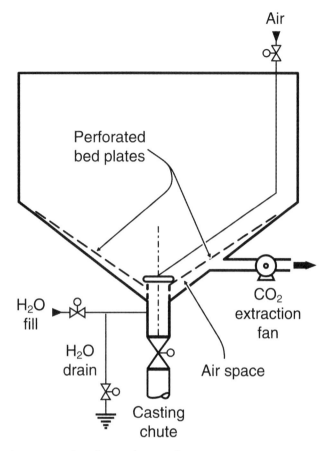

Fig. 21. Diagrammatic representation of a steeping vessel.

The first steep washes a large amount of material off the barley, including dust and leached tannins from the husk. It goes to the drain without reuse and results in a significant effluent charge.

Steeping vessels (Fig. 21) are normally fabricated from stainless steel and most recently have consisted of flat-bottomed, ventilated vessels capable of holding as much as 250 tons of barley. However, the capacity is usually restricted to 50 tons to avoid excessive pressure on grain at the base of the vessel.

A range of process aids has been used to promote the malting operation, generally introduced either in a steep or on transfer from steeping to germination. Potassium bromate (30–200 ppm) may be added to suppress the growth of rootlets within the embryo. Such growth is wasteful and can also cause matting that leads to handling problems. Bromate also suppresses proteolysis. The use of bromate is not permitted in the United States.

Gibberellic acid (GA, Fig. 22) is added in some parts of the world to supplement the native gibberellins of the grain. Although some users of malt prohibit its use, GA can successfully accelerate the malting process. It is usually sprayed onto the grain at levels of 0.1–0.5 ppm as it passes from the last steep on its way to the germination vessel.

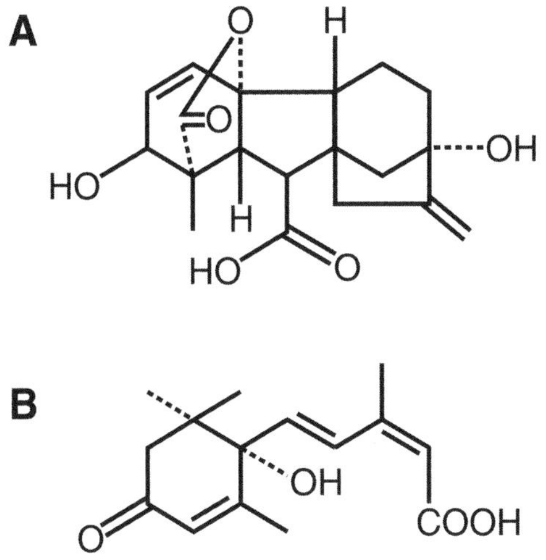

Fig. 22. Chemical structures of some hormones in barley. **A,** gibberellin; **B,** abscisic acid. Gibberellic acid is closely related structurally to the endogenous gibberellins. The synthesis and/or release of the majority of enzymes in the aleurone is promoted by gibberellins but antagonized by abscisic acid, which is not used commercially.

Some maltsters couple the use of GA with a scarification process, whereby the end of the grain furthermost from the embryo is abraded. This enables water and GA to enter the distal end of the grain, triggering enzyme synthesis and modification in the region that is normally the last part of the corn to be degraded. Because these events are also being promoted naturally by the embryo, the resultant effect is called "two-way" modification. It is an opportunity to accelerate the germination process and to deal with barleys that are more difficult to modify.

Some maltsters use hypochlorite or dilute hydroxides to reduce the population of microorganisms on the grain. Such treatments tend to render the malt cleaner and brighter. At one time, maltsters used formaldehyde to extract tannins from the grain and kill microflora, but this practice is virtually unheard of now.

Germination

The hydrated embryo produces hormones that migrate to the aleurone to regulate enzyme synthesis. Principal among these hormones are the gibberellins, which for the most part promote the synthesis of enzymes in the aleurone that break down successively β-glucan in the endosperm walls, protein, and starch.

The gibberellin first reaches the aleurone nearest the embryo, and therefore enzyme release is initially into the proximal endosperm. Breakdown of the endosperm (modification) thus passes in a band from proximal to distal regions of the grain.

There is a view that the scutellum also releases enzymes, but the primary function of that tissue is the absorption and passage to the embryo of the small molecules produced during the degradation of the endosperm.

The rate of modification depends on
- the rate at which moisture distributes through the starchy endosperm
- the rate of synthesis of hydrolytic enzymes

Fig. 23. Floor maltings. (Courtesy of Stan Sole)

- the extent to which these enzymes are released into the starchy endosperm
- the structure (digestibility) of the starchy endosperm.

Factors such as the mealiness or steeliness of grain are important as are the level and organization of molecules such as β-glucan and protein. It is known that certain protein distributions in grain coincide with good maltability and that better malting barleys have less β-glucan and develop more β-glucanase. It is also likely that there is a relationship between maltability and the extent to which β-glucan cross-links with other components in the cell walls. It may also be that better barleys produce more gibberellin and have aleurone cells more responsive to the hormone.

Traditionally, steeped barley was spread out to a depth of up to 10 cm on the floors of long, low buildings and germinated for up to 10 days (Fig. 23). Workers would use rakes either to thin out the grain or pile it up, depending on whether the temperature of the batch needed to be lowered or raised. The aim was to maintain it at 13–16°C. Very few floor malt houses survive because of the intense labor required.

A range of designs of pneumatic (mechanical) germination plants is now used. The earliest such germination vessels were rectangular, fabricated from brick or concrete, and known as Saladin boxes (Figs. 24 and 25). They are still widely used and generally have a capacity of up to 250 tons. The floors of these vessels are made from perforated stainless steel to allow air to pass through the bed of grain. An automated system, such as a helical screw, is used to turn the grain and prevent matting of the rootlets.

Newer germination vessels are circular, made of steel or concrete, can hold as much as 500 tons, and are microprocessor controlled (Figs. 26 and 27). They may incorporate vertical turners located on radial rotating booms, but just as frequently it is the floor itself that rotates against a fixed boom.

Fig. 24. Saladin box. (Courtesy of Stan Sole)

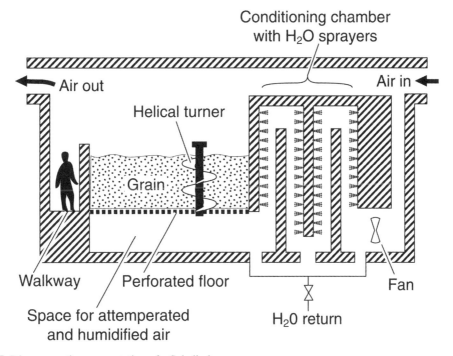

Fig. 25. Diagrammatic representation of a Saladin box.

The modern malting plant is arranged in a tower format, with vessels vertically stacked, the steeping tanks at the top.

Germination in a pneumatic plant is generally at temperatures of 16–20°C. During this process, approximately 4% of the dry weight of the grain is consumed to support the growth of embryonic tissues and much heat is produced. Dissipation of this heat

Fig. 26. Circular germination vessel. (Courtesy of Stan Sole)

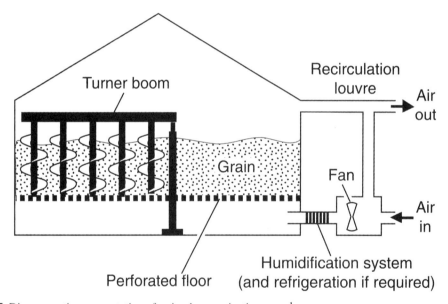

Fig. 27. Diagrammatic representation of a circular germination vessel.

demands the use of large amounts of attemperated air, the oxygen from which is needed by the embryo for respiration. The carbon dioxide produced is flushed away by the air flow.

An experienced maltster spreads a handful of germinating grain on the palm of one hand and then rubs a few corns between the thumb and first finger. If the whole endosperm is readily squeezed out and if the shoot initials (the acrospire, which grows the length of the kernel between the testa and the aleurone and emerges from the husk at the distal end of the corn) are about three-quarters the length of the grain, then the "green malt" is ready for kilning.

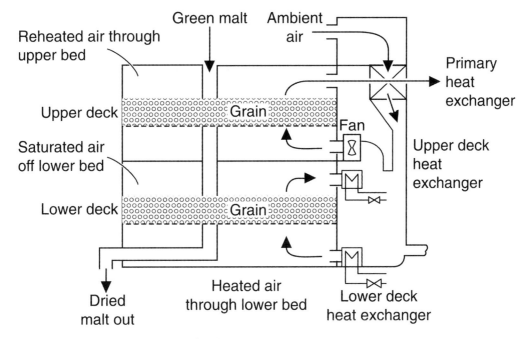

Fig. 28. Diagrammatic representation of a kiln.

Kilning

Through the controlled drying of green malt, the maltster is able to
- arrest modification and render malt stable for storage
- ensure survival of enzymes for mashing
- introduce desirable flavor and color characteristics and eliminate undesirable flavors.

Drying should commence at a relatively low temperature to ensure survival of the most heat-sensitive enzymes (enzymes are more resistant to heat when the moisture content is low). This is followed by a progressive increase in temperature to effect the changes to flavor and color and complete drying within the limited turnaround time available (typically less than 24 hr).

There is a great variety of kiln designs, but most modern ones feature deep beds of malt. They may be rectangular but are more frequently circular in cross section and are likely to be made from corrosion-resistant steel (Fig. 28). They have a source of heat for warming incoming air, a fan to drive or pull the air through the bed, and the necessary loading and stripping systems. The grain is supported on a wedge-wire floor that permits air to pass through the bed, which is likely to be up to 1.2 m deep.

Of course, kilning is an extremely energy-intensive operation, so modern kilns incorporate energy-conservation systems such as glass tube air-to-air heat exchangers. Energy usage has been halved by such systems but can still amount to 2.85 gigajoules per ton of malt.

Newer kilns also use "indirect firing," a process in which the air is warmed by a heater battery containing water as the conducting medium and the products of fuel combustion are exhausted and not allowed to pass through the grain bed. Indirect firing was developed because of concerns that oxides of nitrogen present in kiln gases might promote the formation of nitrosamines in malt (Fig. 29). Nitrosamine levels are now seldom a problem in malt. Maltsters not employing indirect heating systems tend to burn sulfur on the kiln. The sulfur dioxide produced suppresses nitrosamine development, but its acidity does tend to rot the kiln.

There are several phases to the drying process (Fig. 30). At first, there is free drying. Air flows of up to 6,000 m^3 min^{-1} t^{-1} are used. The temperature of the air entering the grain bed (the so-called "air-on" temperature) is 50–60°C. At this stage, the moisture content of the grain drops readily to approximately 23%.

At this point, the water remaining in the grain bed is more resistant to removal and, indeed, is mostly associated with grain components or is not getting to the surface. Because water is not now being easily volatilized and is keeping the temperature of the bed down by latent heat of evaporation, the temperature of the air leaving the kiln ("air-off") starts to rise. This is the "break point."

The air-on temperature is increased and, at this intermediate stage, moisture content in the bed is lowered to 12%. All of the water is now bound, and the temperature is raised again and the fan speed reduced, until the water level in the bed is approximately 6%.

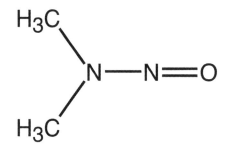

Fig. 29. Chemical structure of nitrosodimethylamine.

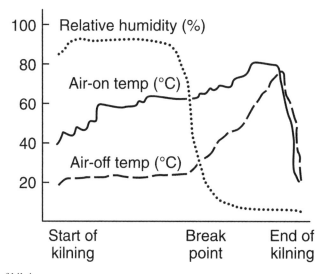

Fig. 30. The stages of kilning.

Finally we reach the "curing" phase, designed to lower the moisture content to final specification, which is typically 4% or less. At this stage, the air-on temperature may be anywhere between 75 and 110°C, depending on the type of malt required. Lower temperatures give malts of lighter color and tend to be employed in the production of malts destined for lager-style beers. Higher temperatures result in darker malts and a wholly different flavor spectrum.

Lager malts result in beers that are relatively rich in sulfur compounds, including dimethyl sulfide. Ale malts have more roast, nutty characters. For both lager and ale malts, kilning is sufficient to eliminate the unpleasant raw, grassy, and beany characters associated with green malt.

When kilning is complete, the heat is switched off and the grain is allowed to cool before it is stripped from the kiln in a stream of air at ambient temperatures. On its way to steel or concrete hopper-bottomed storage silos, the malt is "dressed," which involves mechanical removal of dried rootlets, called "culms" (which go into animal feed), aspiration of dust, sifting out of loose husk and incomplete kernels, and the elimination of any large contaminants.

Types of Malt

It can be seen from the foregoing discussion that numerous variations are possible in respect to malt production, beginning with the barley variety selected and followed by adjustments in malting conditions in terms of out-of-steep moistures, temperature regimes in steeping, germination and kilning, and, where permissible, use of additives such as GA and bromate. Therefore, it is increasingly difficult to generalize about malt types and styles. The following represents no more than a general guideline to malt types; there are ranges of malt types within each category.

Standard malts. *Pilsner malts* are produced from bold (i.e., having big corns) two-row barleys with nitrogen contents of 1.5–1.85%. The barley is steeped to a moisture content of approximately 43% and germinated at relatively low temperatures (below 17°C) to full modification (compared with earlier times, when such malts were relatively poorly modified) before kilning under a slowly rising temperature regime maximizing at less than 85°C. The term "pilsner" is often used unjustifiably for mainstream lager-style beers, whereas it really should be restricted to beers specifically of the type produced in Pilsen in the Czech Republic (if not solely to beers from that part of the world).

Vienna malts are similar to pilsner malts, except that they have a higher color developed through use of barley with slightly greater nitrogen content, are modified to a slightly greater extent, and are kilned at a curing temperature of about 90°C.

Lager malts used in the production of beers more justifiably labeled as "lagers" are produced with protocols similar to those described for pilsner malt, except that the moisture content is likely to be 44–46%.

Munich malts are flavorsome and relatively dark malts produced from high-N barley (e.g., 1.85%), germinated over a long period with subsequent kilning starting at approximately 38°C to allow "stewing" (the continued modification and production of amino acids and sugars, the melanoidin precursors), before progressively raising the temperature through 75°C and 90–100°C to 105°C for final curing.

Pale malts were originally used to produce traditional English-style ales. Barleys should ideally contain less than 1.5% N but may contain up to 1.65%. Barley is steeped to 46% moisture and well modified in germination. Kilning begins at an air-on temperature of about 60°C. The temperature is then raised through 90°C to a final curing temperature of up to 105°C.

Other pale malts used for brewing. *Chit* (or *short-grown*) *malts* are very under-modified and may be used in locations where the addition of unmalted adjuncts is forbidden (e.g., in Germany under the terms of the *Reinheitsgebot*, a law that today lacks rigor under the terms of the European Union but that still has its philosophical adherents). They are cheaper because of their shorter production regimes (e.g., kilning is done after just 1 day of germination). They naturally provide high levels of troublesome cell wall polysaccharides to the wort and in practice can seldom be used at levels above 10% of the grist. They are generally lightly kilned, allowing survival of the relatively low levels of enzymes that have been produced during the short germination time. Chit malts retain some of the unpleasant raw characters associated with barley and green malt.

Green or *lightly kilned malts* in particular still have these raw characters but are very rich sources of enzymes. They are not widely used because of their relatively short shelf life (a result of their high water content).

Diastatic malts are high-enzyme malts valuable for use in mashing with high levels of unmalted adjunct. Because of the higher protein content of six-row barleys, they have a greater potential for producing enzymes. Such barleys are steeped to high moisture contents (47–50%) and germinated under cool conditions for relatively long periods (6–8 days) to enhance final enzyme levels. If permitted, GA is used to further promote enzyme synthesis. Air-on temperatures in kilning may be as low as 35°C, rising to a final temperature not exceeding 55°C to avoid destruction of the enzymes by heat.

Smoked malts are produced by burning peat on the kiln to introduce a smoky, phenolic character. They are used more extensively in the production of whisky than of beer, but with the advent of new beers, particularly by adventurous microbrewers, there is some interest in this style. A long-standing beer that has featured such malt is the German *rauch-bier*, for which the malt is kilned over smoke from burning beech logs.

Wheat malts are used in the production of wheat beers or as adjuncts in ales and stouts. The naked nature of wheat can result in greater damage to the embryo in the malt house; hence, processing must be gentler and vessels may need to be filled to a lesser extent than for barley malts. The grain is steeped to 45% moisture and germinated for a minimum of 5 days at 12–15°C. Kilning is generally fairly gentle, with temperatures often not exceeding 40°C.

Specialty malts. Some malts are produced not for their enzyme content but rather for use by the brewer in relatively small quantities as a source of extra color and distinct types of flavor. They may also be useful sources of natural antioxidant materials. There is much interest in these products for the opportunities they present for brewing new styles of beer. Table 4 describes some of these malts, which are produced in small drum kilns equipped with water sprays (to afford protection against the considerable fire risk). Specialty malts produced with the least extra heat (e.g., cara pils and crystal malt) can be used to introduce relatively sweet, toffee-like characters. Those produced with intense heat (e.g., black malt) deliver pronounced burnt and smoky notes. Issues such as barley

Table 4. Specialty malts

Type	Color (°EBC)	Production Regime
Cara Pils	15–30	The surface moisture is dried off at 50°C before stewing for more than 40 min with the temperature increased to 100°C, followed by curing at 100–120°C for less than 1 hr
Crystal	75–30	As for Cara Pils, but first curing is at 135°C for less than 2 hr
Chocolate	500–1,200	Lager malt is roasted by raising the temperature from 75 to 150°C over 1 hr before allowing temperature to rise to 220°C
Black	1,200–1,400	Similar to chocolate malt, but the roasting is even more intense

variety are not insignificant with regard to specialty malts. Barleys with strong husks are needed to withstand the rigors of the process, which in practice means that plumper two-row varieties tend to be more functional.

Some brewers use non-grist-based sources of color, e.g., caramels. Those laboring under legal restrictions limiting their ability to use such materials have applied considerable ingenuity in their search for ways to adjust their product specifications. *Farbebier* is made from extracts of roast malt that are briefly fermented and charcoal filtered to remove burnt character and is used as a source of color in keeping with the *Reinheitsgebot*. A range of extracts of roasted malts is available in which the color and flavor components have been fractionated and therefore can be used to introduce color without malt flavors and vice versa.

Specifications

The brewer places an increasing number of specifications on malt, parameters that are intended as a guide to how the malt will "behave" in the brewery. The methods employed must be standardized so that the maltster and brewer can have confidence that there will be no discrepancies and therefore no arguments about the quality criteria. In many cases, the methods have been evaluated and recommended by professional bodies, such as the American Society of Brewing Chemists.

The most frequently employed specifications are
- extract, the total amount of material that can be dissolved from malt in a small-scale mash
- moisture content, confirmation that the malt is dried sufficiently and is therefore stable and not "slack"
- modification, which may be evaluated in various ways, including
  - fine/coarse difference: the difference in extract obtained from finely and coarsely milled malt (the smaller the difference the more extensively modified the malt)
  - level of residual β-glucan in the malt, perhaps established by staining of longitudinal sections through the grain with a specific stain called Calcofluor
  - cold water extract, i.e., how much extractable material is already present in malt as opposed to produced during mashing
  - friability, which measures the ease with which the malt can be ground
- nitrogenous components; the total nitrogen content is an inverse measure of starch (i.e., extract). The ratio of nitrogen solubilized in a small-scale mash to the total nitrogen is an index of how extensively broken down is the protein (i.e., protein

modification). The free amino nitrogen is a measure of assimilable amino acids produced from malt.
- fermentability, a measure of the proportion of the solubilized carbohydrates that can be fermented by yeast
- color, confirmation that the requisite degree of kilning has taken place
- enzymes, a measure of the starch-degrading activity (diastatic power) and perhaps residual β-glucan degrading activity (β-glucanase)
- nitrosamines, a safety issue
- flavor components; presently the only one frequently specified is the precursor of dimethyl sulfide (S-methylmethionine)

Further Reading
Briggs, D. E. (1998) Malts and Malting. Blackie Publishing Group, London, U.K.

6. The Components of Barley and Their Degradation During Malting and Mashing

Five separate classes of substance must be considered when discussing the requirements of malting and brewing:
1. cell wall polysaccharides
2. starch
3. protein
4. lipids
5. polyphenols

The cell wall polysaccharides (β-glucans, especially, and pentosans) are problematic because they restrict the yield of extract. They do this either when they are insoluble (by wrapping around the starch components) or when they are solubilized (by restricting the flow of wort from spent grains during wort separation). Solubilized but undegraded β-glucans also increase the viscosity of beer and slow down its filtration. They are also prone to drop out of solution as hazes, precipitates, or gels. Conversely, it has been claimed that β-glucans have positive health attributes, through an ability to lower cholesterol levels and as a contribution to dietary fiber.

Starch is the principal source of fermentable carbohydrate in barley. Failure to properly degrade it not only limits yield, but also presents a clarity problem in beer.

Protein is the origin of the amino acids that are required by yeast for its growth. Some of the protein forms the backbone of beer foam, while other protein (there may be some overlap in the type of proteins involved) is prone to cross-linking with polyphenols to form hazes in beer.

Lipids are detrimental to beer foam and may also be the source of stale flavors in beer.

Polyphenols are key components of hazes that can develop in beers, have antioxidant (protective) functions, and are believed by some to contribute to the mouthfeel of beer.

Cell Wall Polysaccharides

The walls of the endosperm contain virtually no cellulose, but rather are more flexible than cellulosic cell walls (such as those found in the tough husk); this allows their more rapid disintegration during germination.

β-Glucans

β-Glucan molecules are very long (with very high molecular weights) and this, together with the cross-linking capability, means that they are very viscous. If beer containing β-glucan is centrifuged, the glucans tend to unravel; adjacent molecules align with one another and they cross-link through hydrogen bonds between the cellulosic regions (see earlier) to form gels. Agents such as maltose that suppress hydrogen bonding interfere with this interaction, but precipitants such as ethanol promote it. Hence, the risk of gel formation is much greater in beer (especially stronger ones) than in wort.

Apart from cross-links between glucan molecules, they may also cross-link with other types of molecule. This may be significant in the cell wall of barley. There is a suggestion that there are covalent linkages between β-glucan and protein. It is also claimed that the outer region of the cell wall (that which will be encountered first by hydrolytic enzymes produced in the aleurone) is rich in arabinoxylan (pentosan), with the glucan buried within (Fig. 31). There is evidence that the phenolic acids in the cell walls, including ferulic acid, are bound to the pentosan. It seems that it is this type of association that limits the solubility of a proportion of the cell wall polysaccharides (called the "hemicellulose" fraction; the proportion that is freely soluble is called "gum").

The amount of β-glucan in barley depends on variety (malting varieties *may* contain less) but also on environment; there is proportionately more glucan in barley grown under dry conditions.

One of the problems encountered when studying the literature on β-glucans is the lack of uniformity in analysis techniques. A range of procedures has been used to measure β-glucan. None is perfect, but presently two are favored:

- binding of glucan with the agent Calcofluor and measurement of the fluorescence produced. Factors such as pH and temperature have profound effects, and these have seldom been standardized. Also, there is no reaction with glucans below molecular weight 10,000, despite these being potentially problematic.
- hydrolysis of glucan with a β-glucanase enzyme and measurement of the glucose produced.

The enzymic breakdown of β-glucan during the germination of barley and in mashing is in two stages (Fig. 32):

1. solubilization
2. hydrolysis

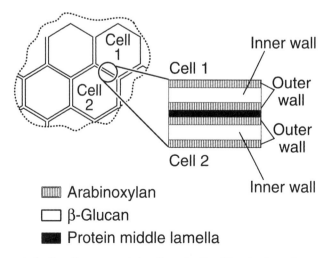

Fig. 31. The arrangement of cell wall components in adjacent cells of the starchy endosperm.

Several enzymes may be involved in the solubilization of glucan. Collectively, the activity is referred to by the trivial name "solubilase." It seems likely that one of the key ways in which glucan can be released is through the breakage of ester bonds binding the glucan into the cell wall. These ester bonds may be between glucan and protein, and the enzyme involved is thought to be an acidic carboxypeptidase. Alternatively, there is evidence for a feruloyl esterase solubilizing glucan and pentosan, which indicates a function for ferulic acid in holding the wall together. Acetylxylan esterase also solubilizes glucan, and so does xylanase, which would be consistent with a masking of glucan by pentosan in the wall.

Much more is known about the hydrolysis stage of glucan degradation. This is catalyzed by endo-β-glucanases (endoenzymes hydrolyze bonds inside a polymeric molecule, releasing smaller units, which are subsequently broken down by exoenzymes that chop off one unit at a time, commencing at one end of the molecule). There are two main endo-β-glucanases in malt that hydrolyze β 1-4 links that are adjacent to a β 1-3 link on the non-reducing side (see Appendix 1 for an explanation of reducing and non-reducing). As a result, these enzymes convert viscous β-glucan molecules to non-viscous oligosaccharides comprising three or four glucose units. Less well understood enzymes are responsible for converting these oligosaccharides to glucose.

Solubilase is present in raw barley and further increases during germination. By contrast, there is little if any β-glucanase in raw barley. It develops during the germination phase of malting in response to gibberellins.

A further differentiation between the enzyme classes is their ability to withstand heat. Solubilase (at least in part) is able to withstand 65°C for 1 hr whereas endo-β-glucanase is destroyed in less than 5 min at this temperature. Furthermore, it is essential that malt is kilned very carefully to conserve β-glucanase activity.

The implication of this difference in heat stability is that solubilase will happily survive mashing, readily releasing into solution any viscous glucan that has survived the malting operation or that is present in any β-glucan-rich adjuncts that may be used (e.g., raw barley, flaked barley, roast barley). The early destruction of the β-glucanase,

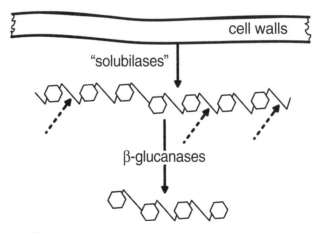

Fig. 32. The degradation of β-glucan.

however, means that there is no capability of dealing with this released glucan. This is why brewers often employ a low temperature start to their mashing processes or traditionally used decoction mashing to deal with less well-modified malts. Alternatively, some brewers add heat-stable β-glucanases of microbial origin.

Strategies to deal with β-glucan commence with the barley selection. Good malting varieties either contain less β-glucan, less problematic β-glucan, or develop more β-glucanase to deal with the glucan.

Most of the degradation of β-glucan occurs during the germination phase of malting. A close relationship exists between the development of β-glucanase and the decline in β-glucan. Factors that increase the development and transport of the enzyme from the aleurone through the endosperm promote cell wall degradation. These factors include (where permitted) the use of gibberellic acid and abrasion and germination at slightly higher temperatures.

It is often overlooked, but modification continues during the early stages of kilning until the temperature becomes too high.

In practice, even the most well-modified of malts contain some residual β-glucan. However, this should present few problems, provided that the malt is homogeneously well modified. The biggest risks arise from the malting of barley that is inhomogeneous and within which modification occurs to different extents (Fig. 33). Reasons for this

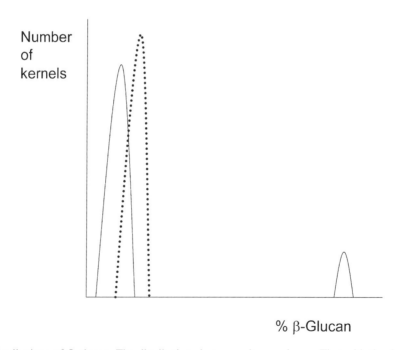

Fig. 33. Distributions of β-glucan. The distributions in two malts are shown. That with the dotted line on average is slightly less well-modified than the other (solid line), as evidenced by the slightly higher residual glucan level. However, the other malt (solid line) contains a proportion of grossly undermodified or unmodified kernels, perhaps as a result of the original barley having an insufficient germinative capacity or germinative energy. These kernels will contribute substantial levels of highly viscous undegraded glucan during mashing.

might include the admixture of different barley varieties in a single batch, barleys that need to be steeped and germinated under different conditions to achieve adequate modification, or the presence of excessive amounts of dead grain (low germinative capacity) or somewhat dormant grain (low germinative energy). Such circumstances are tantamount to having raw barley adjunct as part of the grist.

Malt analysis should include assessment of the homogeneity of modification of malt. This might involve the use of a sanded block technique (i.e., sand the grain down to create a longitudinal section and stain with Calcofluor, which will fluoresce where β-glucan is still present) or even more conveniently by taking the material that does not pass through the sieve on a friabilimeter and subjecting it to a second sieving through a 2.2-mm mesh. By weighing the material collected in the latter fraction, one obtains a count of the grossly undermodified material in a batch of malt.

Pentosans

Alternatively, pentosans are known as the arabinoxylans, because they consist of chains of xylose units (linked β 1-4), with side chains of arabinose linked in α linkages to either C-2 or C-3 of the xylose (or both). Ferulic acid links through ester bonds to the C-5 position of arabinose and accounts for approximately 0.05% of the weight of the wall. It seems that the arabinoxylans in the starchy endosperm cell walls of barley are smaller than the glucans.

There is a lack of understanding about the fine structure of the cell walls of barley, but it is thought by some that the areas of the xylan chain that are not substituted with arabinose may associate through hydrogen bonds with the cellulosic regions on the glucans. There is also a hypothesis that oxidative cross-linking through attached ferulic acids could link adjacent arabinoxylan chains.

At least two enzyme activities are involved in the degradation of the arabinoxylans (in addition to the esterases that remove bound ferulate and acetate). The first (arabinofuranosidase) removes the arabinose side chains, enabling the second (endo-β-xylanase) to chop up the backbone.

It is generally thought, though, that the extent of breakdown of the pentosan in the starchy endosperm is limited during germination. This may be because the necessary enzymes are principally involved in digesting the aleurone walls (which contain largely arabinoxylan, unlike the starchy endosperm walls) and that they are not released into the starchy endosperm until late. It is difficult to rationalize these observations with the more recent ones that suggest that pentosan can limit the accessibility of enzymes to glucan in walls.

Starch

Starch accounts for approximately 65% of the dry weight of barley (Table 5).
The starch in the endosperm of barley is in the form of granules:
- Large granules (15–25 μm in diameter); sometimes referred to as A-type
- Small granules (<10 μm); sometimes referred to as B-type

Both contain some protein (<0.5%) and lipid (<1.2%). This is important because these materials may modify the digestibility of the starch.

50 / Chapter 6

Table 5. Approximate composition of barley

Component	% Dry Weight
Starch	63–65
Protein	10–12
Cell wall polysaccharides	8–10
Sugars	2–3
Lipids	2–3
Lignocellulose (husk)	4–5
Minerals	2

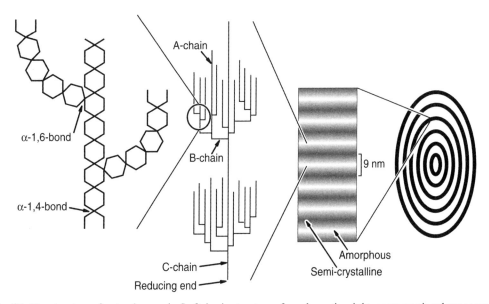

Fig. 34. The structure of a starch granule. Left, basic structure of amylopectin; right, cross-sectional appearance of a starch granule. The relationship between the two is indicated.

The starch in the granules is very highly ordered (Fig. 34), which tends to make the granules difficult to digest. If polarized light is shone through a suspension of granules a "Maltese cross" pattern is observed, because the granules are birefringent (Fig. 35).

When granules are heated (in the case of barley starch beyond 55–65°C), the molecular order in the granules is disrupted (as detected by a loss of birefringence). This process is called *gelatinization*. Now that the interactions (even to the point of crystallinity) within the starch have been broken down, the starch molecules become susceptible to enzymic digestion. It is for the purpose of gelatinization and subsequent enzymic digestion that the mashing process in brewing involves heating.

There is reason to believe that there are small differences in gelatinization temperature in starch from different varieties of barley. This may be caused by different amounts of lipid and/or protein bound to the starch, but it may be due to different packing of the starch molecules themselves.

Any sample of starch gelatinizes over a temperature range rather than suddenly at a single temperature (Fig. 36). This is because a population of granules, even from a single barley, has subtly different packing. What is important in brewing is that the conversion

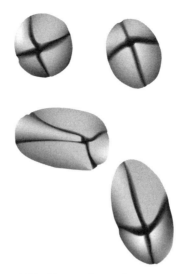

Fig. 35. The "Maltese cross" pattern exhibited by starch granules.

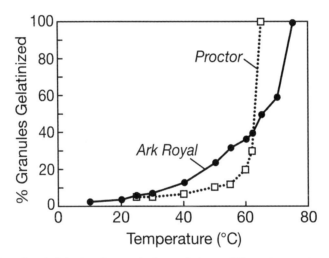

Fig. 36. Extent of starch gelatinization for two barley varieties at different temperatures. (Data from Philip Slack)

temperature in mashing is high enough to gelatinize all of the starch, i.e., is at least the temperature at the top end of the gelatinization range. Thus, while some of the starch granules will gelatinize at 55°C, a typical conversion temperature is 65°C, because many of the granules do not gelatinize until 62°C and some not until the temperature is raised to 65°C.

If granules are maintained in aqueous suspension at a temperature just below their gelatinization temperature, then their gelatinization temperature is raised substantially. This has clear significance because of the range of gelatinization temperatures in granules within one variety and among different varieties. The process is called "granule annealing" and is thought to be due to the "melting" of less crystalline molecules in the granules, which then reconfigure in more crystalline forms.

Another problem caused by heating is the production of "retrograded" starch. If starch is heated at a very high temperature, there may be a tendency for some of the molecules to become resistant to enzymic digestion. Although not documented, it might be expected that such changes would occur during the kilning of malt.

Although 80–90% of the granules in barley are small, they account for only 10–15% of the total weight of starch. The small granules are substantially degraded during the malting process, whereas degradation of the large granules is restricted to a degree of surface pitting. (This is important, as it is not in the interests of the brewer—or maltster— to have excessive loss of starch, which is needed as the source of sugar for fermentation.)

If the small granules survive malting, this is problematic, as they have a higher gelatinization temperature than the large granules and may not be converted under normal mashing conditions. This may not be detected using the iodine test. Iodine gives a strong blue color with the amylose component of starch and this can be used by the brewer to confirm that starch breakdown has occurred in the mash. Small granules contain less amylose than large ones. Small starch granules surviving a mash can complex with other polymers (glucans, oxidized proteins) and contribute insoluble material that retards wort separation.

The starch in barley granules is in two molecular forms (Fig. 37):
- amylose, which comprises linear chains of glucoses joined through α links between C-1 and C-4 of adjacent subunits
- amylopectin, which comprises chains of α 1-4 linked glucose units linked through α 1-6 bonds.

Amylose is the minor component (20–30%). The chain in an amylose molecule tends to be much longer than those in the amylopectin molecule (1,800 glucose units as opposed to 20–25). However, the overall molecular weight of amylopectin is much larger than that of amylose, which means that in any individual amylopectin molecule there must be several thousand chains, not just one as in amylose.

Several enzymes are required for the complete conversion of starch to glucose:
- α-amylase, which is an endoenzyme hydrolyzing α 1-4 bonds within amylose and amylopectin
- β-amylase, which is an exoenzyme, also hydrolyzing α 1-4 bonds, but which approaches the substrate from the non-reducing end, chopping off units of two glucoses (i.e., maltose)
- limit dextrinase, which is an endoenzyme attacking the α 1-6 side chains in amylopectin.

In ungelatinized starch, only the surface of the granule is available for attack, and so digestion is slow (faster for small granules because of the larger surface-to-area ratio— hence, their preferential degradation during germination). Gelatinization exposes the individual starch molecules, and thereafter starch degradation is relatively rapid.

α-Amylase develops during the germination phase of malting. There are several α-amylase isozymes, the most important being complexed with a small protein that inhibits its activity. The inhibitor is cleaved from the enzyme within 15 min at 70°C, i.e., during kilning. This enzyme is extremely heat resistant and also present in very high activity; therefore, it is capable of extensive attack, not only on the starch from malt but also on that from adjuncts added in quantities of 50% or less.

β-Amylase is already present in the starchy endosperm of raw barley in an inactive form through its association with Protein Z. It is released during germination by the action of a proteinase (and perhaps a reducing agent). β-Amylase is considerably more heat labile than α-amylase and is mostly destroyed after 30–45 min of mashing at 65°C.

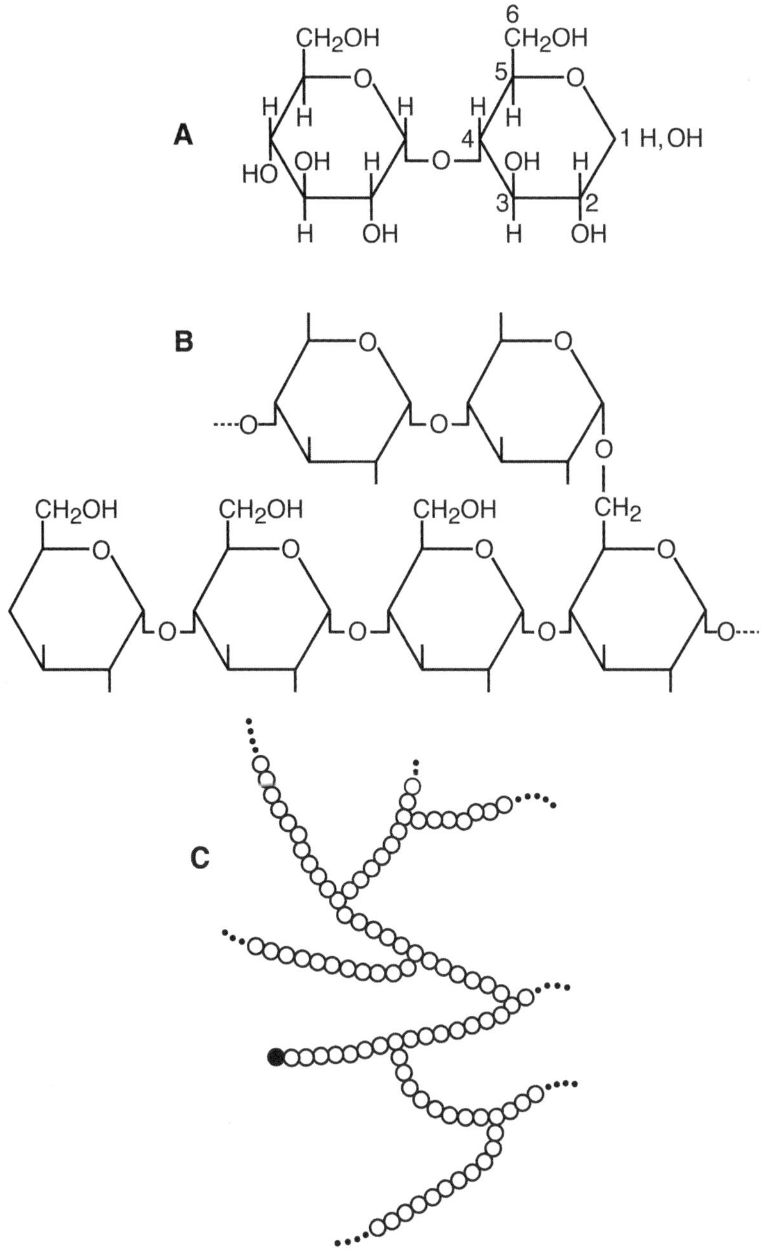

Fig. 37. Linkages in starch (A) maltose repeating unit in amylose and linear portions of amylopectin (B) portion of structure of amylopectin, showing branch point (note: the disaccharide comprising two glucosyls linked through an α 1-6 bond is called isomaltose). Amylopectin (C) with single reducing end as solid circle.

Limit dextrinase is similarly heat-sensitive. Furthermore, it is developed much later than the other two enzymes, and germination must be prolonged if high levels of this enzyme are to be developed. It is present in several forms (free and bound): the bound form is both synthesized and released during germination. This enzyme can be released from its inhibitory binding by lowering mashing pH (e.g., to 5.1) and by reducing conditions.

The β-amylase is incapable of passing a branch point in its hydrolysis of amylopectin. Thus, it is dependent on the action of limit dextrinase if total digestion is to be effected.

In most brew house operations, there is incomplete breakdown of amylopectin because of a combination of circumstances: a high temperature is needed to gelatinize the starch, at which temperature β-amylase and limit dextrinase are being destroyed almost as fast as they can attack the products of α-amylase action. As a result, most worts contain a significant amount of residual dextrin material.

Although it is possible to contrive operations that will allow greater conversion of starch to fermentable sugar, in practice many brewers seeking a more fully fermentable wort add a heat-resistant glucoamylase (e.g., from *Aspergillus*) to the mash (or fermenter). This enzyme has an exo action like β-amylase, but it chops off individual glucose units.

Protein

The proteins in barley have been classified into four groups according to their solubility:
1. albumins, soluble in water (10% of the total protein)
2. globulins, soluble in dilute salt solution (5% of the total protein)
3. hordeins, soluble in alcohol-water mixtures (this fraction is the same as the prolamins in other cereals)
4. glutelins, soluble in dilute acid or alkali or detergent

Hordeins and glutelins comprise 85% of the total protein.

Probably of more value is the division into *storage proteins*—hordeins (especially) and globulins—which serve as reservoirs of nitrogen for the embryo, and *non-storage proteins*—the structural proteins and enzymes.

Hordein, the key storage protein, comprises monomeric and polymeric proteins; in the latter case, they are joined together by disulfide bridges and require the presence of an agent such as mercaptoethanol for their complete extraction. The hordeins are further divided into two major groups (B and C) and two minor (D and γ). Approximately 50% of hordein comprises glutamine and proline.

Polyacrylamide gel electrophoresis (or other protein separation and visualization procedures) can separate extracted hordeins, the resultant pattern being useful as an aid to identifying barley varieties and indicating which ones may have good malting and brewing potential. The pattern does not change with environmental conditions and is retained (despite proteolysis) in the early stages of germination.

There is still dispute about the relative significance of malting and brewing as stages in protein degradation. It is now generally thought that germination is the most important stage for degradation of the proteins into smaller units (peptides and amino acids). There may be some ongoing protein extraction and precipitation during mashing, and smaller peptides may be converted into amino acids at this stage, but wholesale proteolytic degradation in mashing is thought to be limited.

Two categories of enzyme are responsible for protein degradation. The endopeptidases (proteases) generate peptides by attacking at the heart of the protein molecules, with these peptides being subsequently hydrolyzed by an exoenzyme, carboxypeptidase, that chops off individual amino acids sequentially from the carboxyl end of the peptides.

The endopeptidases are synthesized during germination in response to gibberellin. There are at least 40 such enzymes. Proteolytic inhibitors are also present in the grain, and these may be progressively separated from the proteases during germination. The proteases are relatively heat-labile (like the endo-β-glucanases). One of the likely reasons for limited proteolysis during mashing is that the endogenous proteinase inhibitors are extracted from the milled grain and are now free to block the enzymes from which they were kept separate in the intact kernel.

Substantial carboxypeptidase is present in raw barley (it may also act as a β-glucan solubilase, see earlier) and it further increases in amount during germination. It is quite heat-resistant and (like α-amylase) is unlikely to be limiting. Thus, the extent of protein degradation is largely a function of the extent of protease activity during germination.

Lipids

The lipids can be divided into the starch lipids (those associated with the starch granules) and the non-starch lipids. In barley, the order of magnitude is 75% non-polar lipid (e.g., glycerides), 10% glycolipid, and 15% phospholipid.

Perhaps a third of the lipid is present in the embryo, with the remainder being located in the endosperm (starchy endosperm plus aleurone).

The lipids are significant because they influence yeast action in fermentation (and therefore the beer quality, including flavor), but also because they are detrimental to foam stability and flavor life. In the latter context, the most significant component of the lipid is the fatty acid, an approximate distribution for which is

linoleic acid	(C18:2)	58%
palmitic acid	(C16:0)	20%
oleic acid	(C18:1)	13%
linolenic acid	(C18:3)	8%
stearic acid	(C18:0)	1%

Of especial concern are the polyunsaturated fatty acids (linoleic and linolenic acids) that are susceptible to oxidation, leading to rancidity and staling.

Lipids, by definition, have limited solubility in water. For this reason physical effects are generally more important than chemical or biochemical ones in brewing. The lipids (certainly if undegraded) tend to associate with insoluble components of the mash, and therefore they are largely lost with the spent grains and become associated with hot and cold breaks.

There are, however, enzymes present in malt that are capable of hydrolyzing lipids. The prime enzyme of attack is lipase, which splits fatty acids from glycerol. There is probably a range of lipases that act on different types of lipid (non-polar, phospho-, glyco-). They have been poorly characterized.

Much more attention has been lavished on lipoxygenase, which is capable of oxidizing unsaturated fatty acids (linoleic acid and linolenic acid) to hydroperoxides. These products can then be converted into staling aldehydes.

Barley develops two lipoxygenase enzymes in the embryo. Both enzymes are extremely heat sensitive and are extensively lost during most kilning regimes. Extracts contain variable levels of endogenous inhibitors (chiefly polyphenols), which need to be dialyzed away before full activity can be assayed.

The level of lipoxygenase present in malt declines during storage, and it has been hypothesized that this is the reason that malt should be stored for a few weeks prior to use. The enzyme is involved in oxidative reactions that contribute to the formation of the

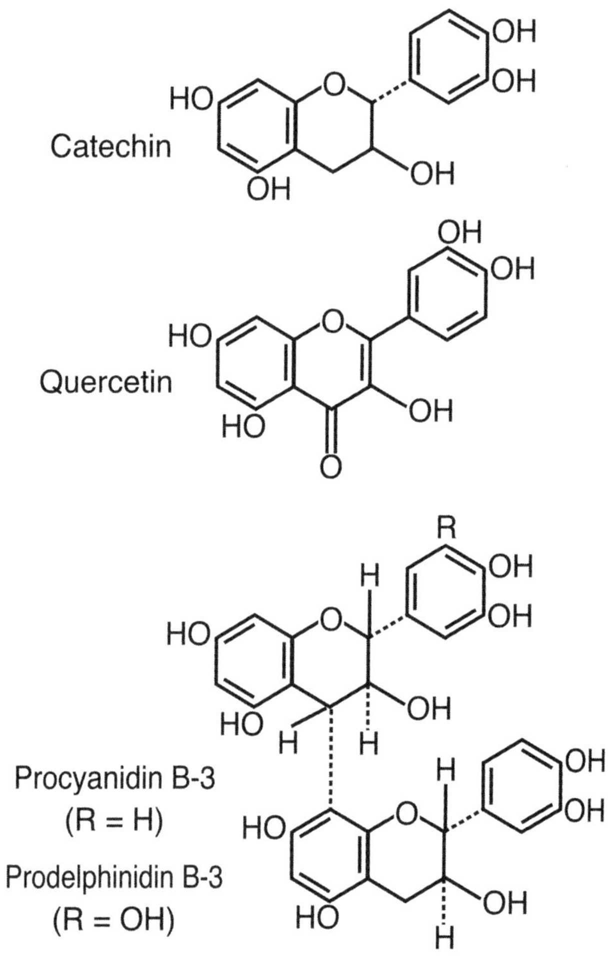

Fig. 38. Some polyphenols.

oberteig layer in the grains bed (see later) that impedes wort separation; and when the level of the enzyme reduces, so too does the ease of wort separation improve.

Polyphenols

Barley (and hops) contains significant quantities of phenolic materials. We have already encountered ferulic acid as a component of the cell walls in the endosperm. Quantitatively of more significance are the polyphenols, which have a very complex chemistry (Fig. 38). Polyphenols such as catechin and epicatechin are capable of entering into oxidation-reduction reactions. Their ability to be oxidized is beneficial from the aspect of them acting as antioxidants. However, when they are oxidized they tend to polymerize and cross-react to form precipitates with proteins. If this occurs in beer, then the result is a haze.

Although polyphenols can oxidize in the absence of an enzyme, this process is greatly accelerated by enzymes. Raw barley contains polyphenol oxidase (laccase), but this is almost entirely lost during malting. Of much more significance in a mash is peroxidase. This will react polyphenol with hydrogen peroxide (formed by the reaction of thiol groups in gel proteins with oxygen) to form reddish complexes and promote insolubilization. There are a large number of heat-stable isoforms of peroxidase in malt, and their significance has probably been underestimated over the years.

Further Reading

MacGregor, A. W., and Bhatty, R. S., eds. (1993) Barley: Chemistry and Technology. American Association of Cereal Chemists, St. Paul, MN.

7. Production of Sweet Wort

Sweet wort is the sugary liquid that is extracted from malt (and other solid adjuncts used at this stage) through the processes of milling, mashing, and wort separation.

Key issues at this stage in beer production are
- yield of extract per brew
- numbers of brews that can be achieved per day
- quality of wort in terms of desired fermentability, perhaps clarity, and perhaps extent of oxidation.

Malt and Solid Adjunct Handling

Larger breweries have raw materials delivered in bulk (by rail or road). Railcars may carry up to 80 tons of malt and a truck 20 tons. Smaller breweries have malt delivered by sack. The delivery and the vehicle it came in are checked for cleanliness and sampled with a trier. The sample will be inspected visually and smelled before unloading is permitted. Most breweries spot check malt deliveries for key analytical parameters against the agreed contractual specification.

Grist materials are stored in silos sized according to brew house throughput. A mash size (called "brew length") in a large brewery might be 1,000 hl. At a water-to-grist ratio of 3:1 (3 parts water to 1 part malt), this amounts to over 30 tons of malt.

Milling

Before malt or other grains can be extracted, they must be milled after a cleaning and screening operation to remove dust and debris (Figs. 39 and 40) and via weighing to ensure delivery of the appropriate mass of grain (Fig. 41).

The more extensive the milling, the greater the potential there is to extract materials from the grain. However, in most systems for separating wort from spent grains after mashing, the husk is important as a filter medium: the more intact the husk, the better the filtration. Therefore, milling must be a compromise between thoroughly grinding the endosperm while leaving the husk as intact as possible.

There are fundamentally two types of milling:
1. dry milling
2. wet milling

Dry milling is more common.

Dry Milling

Mills may be roll, disk, or hammer. If wort separation is by a lauter tun, then a roll mill is used. If a mash filter is installed, then a hammer (or disk) mill may be em-

60 / Chapter 7

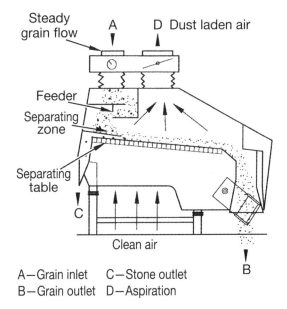

Fig. 39. Destoner. (Courtesy of Briggs of Burton)

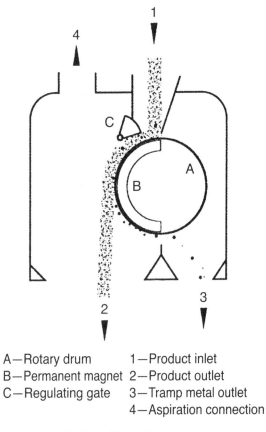

Fig. 40. Magnetic separator. (Courtesy of Briggs of Burton)

ployed. This is because the husk is much less important for wort separation by a mash filter.

Mill settings are modified (regularly but not continually) in response to the quality of the grist. A finer grind is needed for malt of lower modification. The brewer will inspect the milled grist by using a sieve analysis. A comparison of the grist analysis that might be used for the different types of brew house is shown in Table 6.

Fine milling permits a greater extraction of undesirable as well as desirable components from the grist. Thus, lipids, polyphenols, and residual β-glucans are more readily extracted.

There are several types of roll mill: basically more rolls enable greater flexibility.

A two-roll mill is satisfactory for dealing with well-modified malt. In such a mill the malt simply passes through two rolls set such that the gap within them crushes the kernels.

In four- and six-roll mills (Fig. 42), there are two and three pairs of rolls set successively more closely together (e.g., 1.3–1.5 mm, 0.7–0.9 mm, 0.3–0.35 mm). The

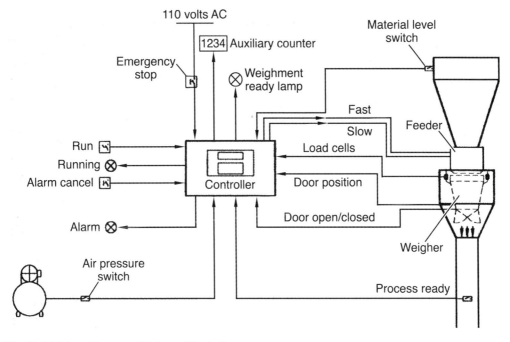

Fig. 41. Weigher. (Courtesy of Briggs of Burton)

Table 6. Milled grist distributions for different wort separation systems

Sieve Mesh (mm)	What is Retained	% Needed for Lauter Operation	% Needed for Mash Filter Operation
1.27	Husk	18	11
1.01	Coarse grits	8	4
0.547	Fine grits 1	35	16
0.253	Fine grits 2	21	43
0.152	Fine grits 3	7	10
Residue	Powder	11	16

62 / Chapter 7

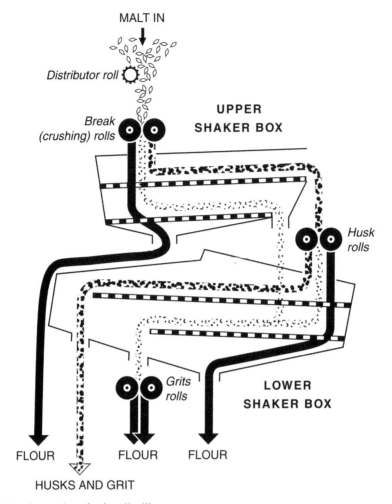

Fig. 42. Principle of operation of a six-roll mill.

rolls may have a diameter of 30 cm and a length between 45 and 150 cm, and they may be grooved or fluted. There may be shaker boxes between the rolls to facilitate operation.

These mills are able to deal with less well-modified malts. Flour produced by the first pair of rolls falls through to the grist case, and fine grits are diverted to the third pair of rolls. The second pair of rolls crushes the hard ends (the relatively undermodified distal parts of the malt) and coarse grits are reduced in size. Again flour passes to the grist case and remaining grits pass to the third rollers. Depending on the size of the rolls, a mill with 150-cm-long rolls might process as much as 5 tons of malt per hour.

The two- and four-roll mills are suitable for grinding broken rice. Barley, wheat, and whole grain rice may be milled with a six-roll mill with modified set-up.

Hammer mills comprise a horizontal rotating spindle to which are attached a series of hammers, each of which is attached to the spindle through a pivot (Fig. 43). The spindle is enclosed in a casing. Between the spindle and the casing at the base is a sieve.

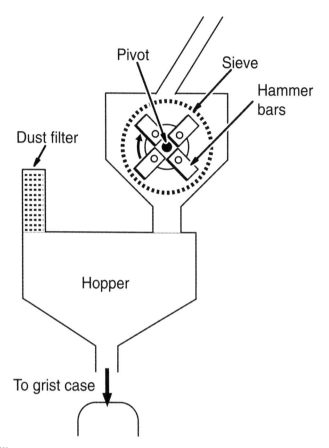

Fig. 43. Hammer mill.

Wet Milling

Wet milling was introduced into some brewing operations as an opportunity to minimize damage to the husk on milling and to lessen dust risks.

At one extreme, some brewers expose dry malt to steam prior to dry milling: the steam moistens the husk (but not the endosperm, which if wet will stick to the rolls) and renders it less prone to disintegration in the mill. This technique is called "conditioning."

Wet milling proper is illustrated in Figure 44. Malt is steeped in water (25–50°C) for about 15 min, allowing the moisture content of the grain to increase to up to 30%. The water is then drained off to the mashing vessel (because it will contain some extract), and the malt is delivered by a feed roll to the two-roll mill. The rolls are lightly patterned but not grooved. From there the crushed malt drops to a "splasher plate" that is rotated by a centrifugal pump to intimately mix the grist with mashing-in water. From there the mixture is pumped to the mashing vessel.

Wet milling is constraining insofar as it inextricably ties the milling in to the mashing, and the opportunity for oxygen pick-up during pumping is thought by many to be detrimental to quality.

Milling operations (especially when dry) constitute an explosion hazard, and therefore dust prevention and removal precautions and efficient cleaning are essential.

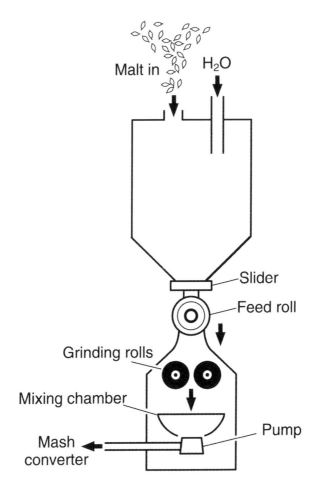

Fig. 44. Wet mill.

The Grist Case

Dry milled "goods" are collected from the mill in a grist case (or hopper). Hoppers need to be quite large, because the density of ground malt is very low.

A single grist case may service one or more mashing vessels and may serve the purpose of mixing the ground malt with other solid adjuncts, either after milling or if they arrive in a flaked form. Alternatively, there may be separate malt and adjunct hoppers.

Residence time in the case is usually short, but it does afford a buffer facility in case of delays. Hoppers, generally fabricated from mild steel, are on load cells to facilitate accurate computation and delivery of grist according to its weight.

Mashing

Mashing is the process of mixing milled grist with heated water in order to digest the key components of the malt and generate wort containing all the necessary ingredients for the desired fermentation and aspects of beer quality.

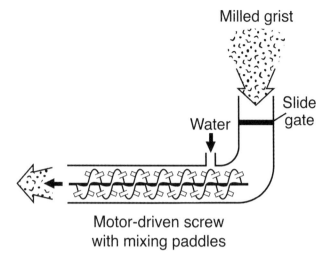

Fig. 45. Steel's masher.

There are several types of mashing:
- infusion mashing
- decoction mashing
- temperature-programmed mashing

Whichever type of mashing is employed, the vessels nowadays are most frequently fabricated from stainless steel (once they were copper). What stainless steel loses in heat transfer properties is made up for in its toughness and ability to be cleaned thoroughly by caustic and acidic detergents. (Some brewers still have copper vessels, but more often the copper is simply an external coating to a stainless steel vessel, the copper being considered more attractive.)

Most mashing systems (except for wet milling operations) incorporate a device for mixing the milled grist with the water (which some brewers call "liquor"). This device, the "pre-masher," can be of various designs, the classic one being the Steel's masher, which was developed for the traditional infusion mash tun (Fig. 45).

The Steel's masher comprises a screw-drive motor driving a series of mixing paddles that blend milled grist with water en route to the mash tun. In traditional infusion mashing, this is the only opportunity for mixing. Modern mashing vessels incorporate agitators (rousers) but nonetheless still benefit from the use of a pre-masher. This may be nothing more sophisticated than a chamber where passing grist is sprayed with water.

Infusion Mashing

Infusion mashing is relatively uncommon but still championed by traditional brewers of ales. It was designed in England to deal with well-modified ale malts that did not require a low-temperature start to mashing in order to deal with residual cell wall material (β-glucans).

Grist is mixed with water in a Steel's masher en route to the preheated mash tun (Fig. 46). A single holding temperature, classically 65°C, is employed. This temperature facilitates gelatinization of starch and subsequent amylolytic action.

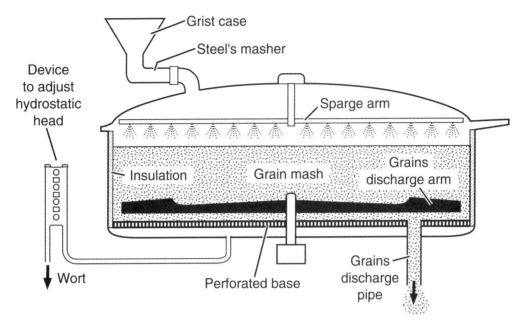

Fig. 46. Mash tun.

At the completion of this "conversion," wort is separated from the spent grains in the same vessel, which incorporates a false bottom and facility to regulate the hydrostatic pressure across the grains bed. The grist is sparged to enable leaching of as much extract as possible from the bed.

Decoction Mashing

Decoction mashing was developed in Europe to deal with lager malts that were less well-modified than ale malts and in days before temperature could be measured accurately. Essentially, it allows the facility to start mashing at a relatively low temperature, thereby allowing hydrolysis of the β-glucans present in the malt and then raising the temperature to a level sufficient to allow gelatinization of starch and its subsequent enzymic hydrolysis.

The temperature increase is achieved by transferring a portion of the initial mash to a separate vessel, where it is taken to boiling and then returned to the main mash, leading to an increase in temperature. This is a rather simplified version of the process, which traditionally involved several steps of progressive temperature increase (Fig. 47).

Temperature-Programmed Mashing

Although there are some adherents to the decoction mashing protocol, most brewers today employ the related but simpler temperature-programmed mashing. Again, the mashing is commenced at a relatively low temperature, but subsequent increases in temperature are effected in a single vessel by employing steam-heated jackets around the vessel to raise the temperature of the contents, which are thoroughly mixed to ensure even heat transfer. A diagram of a modern mashing vessel is given in Figure 48.

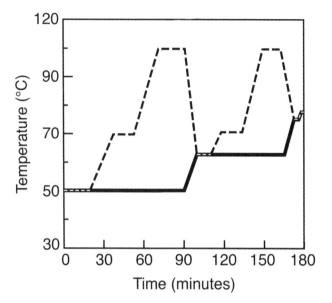

Fig. 47. Decoction mashing regime. --- Boiled mash; — main mash. A short start (here, 20 minutes) at (here, 50°C) is followed by removing a proportion of the mash, taking it to boiling, and adding it back to the main mash to raise the temperature to 65°C. The subsequent second "decoction" serves to raise the temperature to that required for wort separation.

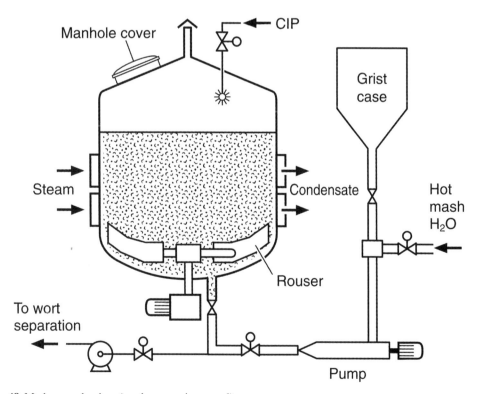

Fig. 48. Modern mash mixer (mash conversion vessel).

Mashing may commence at 45–50°C, followed by a temperature rise of 1 degree C per minute until the conversion temperature (63–68°C) is reached. The mash will be held for perhaps 50 min to 1 hr before the temperature is again raised to the sparging temperature (76–78°C). High temperatures are employed at the end of the process to arrest enzymic activity, facilitate solubilization of materials, and reduce viscosity, thereby allowing more rapid liquid-solid separation.

Accurate temperature and mass/volume control are essential for efficient control of a mashing operation. The temperature is monitored by probes, the mass of grist by load cells, and the volume of water by meters.

Dealing with Adjuncts

Whether to use an adjunct is decided on the basis of

- cost: does it represent a cost-advantageous source of extract compared with malted barley? (That is, is the reduced cost of an adjunct still worthwhile after any extra expenditure associated with using it is subtracted?)
- quality: does the adjunct provide a quality benefit in respect to flavor, foam, color, etc.?

Liquid adjuncts (sugars and syrups) are added during the wort boiling stage. A series of solid adjuncts may be added at the mashing stage, because they depend on the enzymes from malt to digest their component macromolecules.

Solid adjuncts may be based on cereals other than barley, such as wheat, corn (maize), rice, oats, rye, or sorghum. These adjuncts can be in different forms:

- raw cereal (barley, wheat, rice)
- raw grits (corn, rice, sorghum)
- flaked (corn, rice, barley, oats)
- micronized or torrefied (corn, barley, wheat)
- flour/starch (corn, wheat, sorghum)
- malted (wheat, oats, rye, sorghum)

A key aspect of solid adjuncts is the gelatinization temperature of the starch (Table 7). The higher gelatinization temperature for corn, rice, and sorghum mean that these cereals need treatment at higher temperatures than do barley, oats, rye, or wheat. If the cereal is in the form of grits (produced by the dry milling of cereal to remove outer layers and the oil-rich germ), then it needs to be "cooked" in the brew house.

Alternatively, the cereal can be preprocessed by intense heat treatment in a micronization or a flaking operation. In the first process, the whole grain is passed by conveyor

Table 7. Gelatinization temperatures of starches from different cereals

Source	Gelatinization Temperature (°C)
Barley	61–62
Corn	70–80
Oats	55–60
Rice	70–80
Rye	60–65
Sorghum	70–80
Wheat	52–54

under an intense heat source (260°C), resulting in a "popping" of the kernels (cf., puffed breakfast cereals). In flaking, grits are gelatinized by steam and then rolled between steam-heated rollers. Flakes do not need to be milled in the brew house, but micronized (torrefied) cereal does.

Cereal cookers employed for dealing with grits are made of stainless steel and incorporate an agitator and steam jackets. The adjunct is delivered from a hopper and mixed with water at a rate of perhaps 15 kg per hectoliter of water. The adjunct is mixed with 10–20% of malt as a source of enzymes.

Following cooking the adjunct mash is likely to be taken to boiling and then mixed with the main mash (at its mashing-in temperature), with the resultant effect being the temperature rise to conversion for the malt starch (cf., decoction mashing). This is sometimes called "double mashing."

Wort Separation

Recovering wort from the residual grains in the brewery is perhaps the most skilled part of brewing. Not only is the aim to produce a wort with as much extract as possible, but many brewers prefer to do this such that the wort is "bright," i.e., not containing many insoluble particles that might present difficulties later. All this needs to take place within a time window, for the mashing vessel must be emptied in readiness for the next brew.

Irrespective of the system employed for mash separation, Darcy's equation applies:

$$\text{rate of liquid flow} = \frac{\text{pressure} \times \text{bed permeability} \times \text{filtration area}}{\text{bed depth} \times \text{wort viscosity}}$$

And so the wort will be recovered more quickly if the device used to separate the wort has a large surface area, is shallow, and if a high pressure can be employed to force the liquid through.

The liquid should be of as low viscosity as possible, as less viscous liquids flow more readily.

Also the bed of solids should be as permeable as possible. Perhaps the best analogy here is to sand and clay. Sand comprises relatively large particles around which a liquid will flow readily. To pass through the much smaller particles of clay, though, water has to take a much more circuitous route and it is held up.

$$\text{Permeability} = \frac{\text{bed porosity}^3 \times \text{particle diameter}^2}{180 \times (1 - \text{bed porosity})^2}$$

$$\text{Bed porosity} = \frac{\text{wort volume}}{\text{bed volume}}$$

The particle sizes in a bed of grains depend on certain factors, such as the fineness of the original milling and the extent to which the husk survived milling (see earlier). Furthermore, a layer (*teig* or *oberteig*) collects on the surface of a mash. This layer consists of a complex of certain macromolecules, including oxidatively cross-linked proteins, lipids, cell wall polysaccharides, and any surviving small starch granules and has a very fine size distribution analogous to that of clay.

However, particle size also depends on the temperature, and it is known that at the higher temperatures used for wort separation (e.g., 78°C) there is an agglomeration of very fine particles into larger ones through which wort will flow more quickly.

Lauter Tun

Generally, a lauter tun is a straight-sided cylindrical vessel with a slotted or wedged wire base and run-off pipes through which the wort is recovered (Fig. 49). Within the vessel there are arms that can be rotated about a central axis. These arms carry vertical knives that are used as appropriate to slice through the grains bed and facilitate run off of the wort.

The brewer first runs hot liquor (at about 77°C) into the vessel such that it rises to an inch or so above the false bottom. This ensures that no air is trapped under the plates, and it also serves to "cushion" the mash. The mash is then transferred carefully from the mash tun to the bottom of the vessel, again to minimize oxygen uptake, and the knives are used to ensure that the bed is even.

Hot water is used to "rinse out" the mash tun and delivery pipes. The depth of the grain bed is unlikely to be more than 18 in.

After a "rest" of perhaps 30 min, the initial stage is to run off from the base of the vessel and recycle this wort into the vessel, in order that it can be clarified. After 10 to 20 min of this so-called "vorlauf" process, the wort is diverted to the kettle and wort collection proper is started. This wort is at its most concentrated.

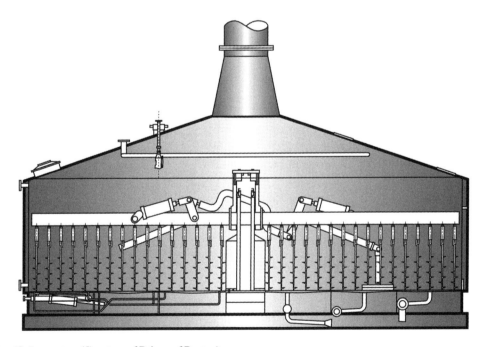

Fig. 49. Lauter tun. (Courtesy of Briggs of Burton)

The remainder of the process is an exercise in running off as concentrated a wort as possible within the timeframe available. Hotter (77°C) liquor (the "sparge") is sprayed onto the grains in order that the sugars and other dissolved materials are not left trapped in the spent grains. The knives are used as sparingly and carefully as possible so as not to damage grains and thereby make small particles that would "clog" the system or render the wort turbid or "dirty."

Another factor that the brewer must consider is the strength of wort that is needed in the fermenter. If the intention is to brew a very strong beer, then clearly the wort must be rich in sugars. This limits the amount of sparge liquor that can be used in the lauter tun. Some brewers collect the initial stronger worts running off from the lauter tun separately in one kettle, using this for stronger brews, before collecting subsequent weaker worts in a second kettle ("parti-gyling").

When the kettle is full, there may still be some wort left with the grains. Time permitting, this is run off for use as "mashing-in" liquor for subsequent brews, a process referred to as "weak wort recycling." The brewer needs to be careful: when the worts are very weak and have less buffering capacity (pH is high) there is an increased tendency to extract tannins out of the grains, and these can cause clarity problems in beer.

At the completion of lautering, grain-out doors in the base of the vessel are opened, and the cutting machinery is used to drive the grains out. Almost without exception, spent grains are trucked off site as fast as possible (they readily "spoil") for direct use as cattle feed. Drying is an extremely expensive alternative but may be forced upon the brewer in some circumstances.

Mash Filters

Increasingly, modern breweries are using mash filters (Figs. 50 and 51). These operate by using plates of polypropylene to filter the liquid wort from the residual grains. Accordingly, the grains serve no purpose as a filter medium and their particle sizes are irrelevant. The high pressures that can be used overcome the reduced permeability due to smaller particle sizes. Furthermore, the bed depth is particularly shallow (2–3 in.), being nothing more than the distance between the adjacent plates.

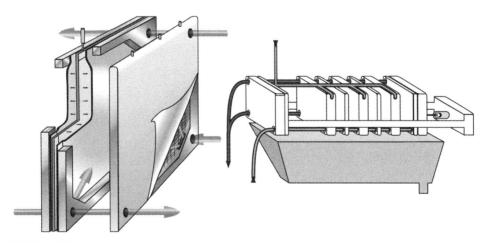

Fig. 50. Mash filter.

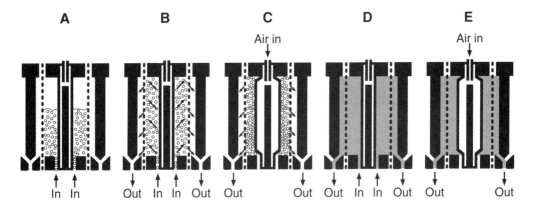

Fig. 51. Principle of operation of a mash filter. A, filing; B, filtration; C, precompression; D, sparging; E, compression.

The chambers of the press are first filled with water, which is then replaced by mash with filling times of less than 30 min. During this time the first worts are recovered through the plates. Once full, the outlet valves are closed. The filter is then given a gentle compression to collect more wort. This is followed by sparging to get a uniform distribution of liquor across the filter bed and then a further compression to force out the remaining wort.

Using mash filters, wort separation can be completed in 50 min rather than the periods of up to 2 hr needed for lautering.

Relating the Chemistry and Biochemistry of Mashing to Brew House Operations

Parameters such as hot-water extract are derived from small-scale mashes. These are a reasonable but imperfect mimic of a large-scale brewing operation. The realization of high yields of extract with an avoidance of problems in the brew house (e.g., slow rates of wort separation) and an achievement of the necessary quality criteria in the wort are as dependent on the operations in the brew house as they are on the raw materials entering into that brew house. Both need to be right.

The mashing operation involves several events:
- leaching of preformed soluble materials from the grist
- extraction of substrates from the grist and enzymes from the malt
- enzyme action
- chemical interactions between mash components

Cell wall components. Most cell wall components should be removed in the germination and early kilning phases of malting. However there will always be some residual β-glucan and pentosan, especially if adjuncts are used that are rich in these materials (barley- and wheat-based adjuncts, respectively). If there is a glucan-rich grist then it is usually necessary to commence mashing at a relatively low temperature to allow action from the heat-sensitive endo-β-glucanase. Regimes such as decoction and temperature-programmed enable this to occur. Alternatively a heat-tolerant microbial β-glucanase might be employed.

Table 8. Malt and wort carbohydrates (% of total)

	Malt	Wort
Starch	85.5	Negligible
Dextrin	Negligible	22
Maltotetraose	Negligible	6
Maltotriose	<1	14
Maltose	1	41
Sucrose	5	6
Glucose + fructose	2.5	9
Other	5	2

Starch. Table 8 gives a broad comparison of the carbohydrate distribution (%) in malt and in the resultant wort extracted at 65°C.

The precise distribution of carbohydrates in the wort will depend on the grist and on the mashing conditions. Of particular significance are the levels of various amylases present in the grist, the temperature, and the mash thickness.

α-Amylase is present in considerable quantities and it is a relatively heat-tolerant enzyme. This enables it to "dextrinize" the starch not only of the malt but also of significant quantities of adjunct (after gelatinization). The term dextrinize refers to the primary action of α-amylase in converting starch into shorter chains of glucose units called dextrins.

By contrast, β-amylase and limit dextrinase are less stable and less prevalent. Thus if a malt is "diluted" with other grist components or if a more highly kilned malt is used, then there will be proportionately more α- than β-amylase and the yield of fermentable extract will decline (β-amylase is the key enzyme for converting dextrins into a fermentable form).

Furthermore, if the conversion temperature is progressively increased, the imbalance between the α-amylase and the other two enzymes also becomes exaggerated and fermentability declines. Some brewers take advantage of this in the making of less-fermentable worts for the production of low-alcohol beers; they may employ a conversion temperature of 72–74°C.

Generally, a brewer adds exogenous heat-stable glucoamylase if there is a need to maximize fermentability, e.g., in the production of "diet" beers. However, there are alternative strategies, such as the extensive use of decoction procedures (the mash component being boiled achieves gelatinization and when it is returned to the main mash there is then opportunity for considerable action by α- and β-amylases and limit dextrinase). Another approach might be to add an extract of lightly kilned malt to a fermenter.

Protein. Earlier, we stressed that protein degradation is largely a function of the malting process. Although the term "protein rest" has long since been adopted for the low-temperature commencement of many mashing operations, in fairness this would be much more accurately called a "glucanolytic rest."

Effects of other mashing variables. The brewer can adjust certain other components of the mash in order to influence mashing and wort composition. These are salt levels, pH, and mash thickness.

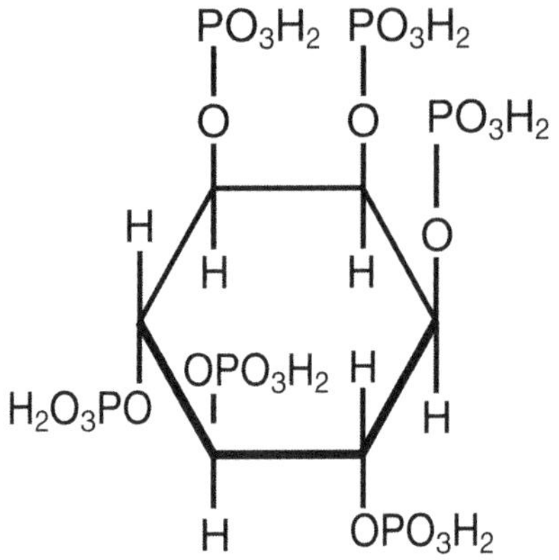

Fig. 52. Phytic acid.

We will look into water composition later. For now, we need consider only calcium and phosphate. Both are derived to some extent from malt as well (in the case of calcium) as water. Malt contains both inorganic phosphate and organic phosphate in the form of phytate (inositol phosphate, Fig. 52).

Calcium is a good precipitant. In particular, it reacts with phosphate to lower the pH and with oxalic acid. Failure to remove oxalic acid at this stage leads to the risk of its precipitating in beer to cause gushing and haze.

pH. pH is important in mashing because enzymes are strongly affected by pH. Furthermore, pH influences the solubility and extractability of molecules.

The pH of a mash depends on mash temperature, and at 65°C the pH of a mash is approximately 0.35 less than at 18°C because of dissociation of buffer substances. This difference must be considered when measuring pH and when declaring the "optimum pH" for mashing events.

Although the precise picture will depend on the nature of the grist and the mashing regimes involved, it has been shown that the highest yield of extract and highest fermentability occur at pH 5.3–5.8 (measured at 65°C). The highest yields of soluble nitrogen and free amino nitrogen, however, occur at a lower pH of 4.7–5.2.

The pH of a mash depends on the buffering potential of the system, but an all-malt mash made with deionized water has a pH of around 5.8 (although this will depend on the malt; increased modification and greater color mean lower pH).

Addition of calcium ("Burtonization") lowers the pH, as will the addition of acid (e.g., phosphoric).

In Germany, under the terms of the *Reinheitsgebot*, acidification must be natural and is achieved by deliberate infection of mash with *Lactobacillus delbruckii* or the addition to a mash of acid produced in the brewery in a separate lactic fermentation. (An alternative is to use malt steeped in lactic acid during its production.)

Mash thickness. The thicker the mash, the greater the survival of heat-labile enzymes, as the particles afford protection. However, thinner mashes can be mixed more readily, and extraction of materials is more efficient. Furthermore, product feedback inhibition of enzymes is reduced.

Table 9. Values required for computation of striking temperature

Moisture Content of the Malt (%)	Specific Heat of the Malt Grist	Slaking Heat Correction (°C)	Slaking Heat Correction (°F)
0	0.38	3.1	5.5
1	0.38	2.6	4.7
2	0.39	2.3	4.1
3	0.40	2.1	3.7
4	0.40	1.7	3.1
6	0.41	1.3	2.3
8	0.42	1.1	2.0

Calculations

The brewer needs to calculate how much malt (and adjunct) is needed to generate a certain amount of wort of a certain strength. Such calculations can be only approximations insofar as the extract is measured on a malt or adjunct in a small-scale mash and various factors will dictate how much extract is realized on a large scale (e.g., wort run-off difficulties).

Example

You require 800 hl of wort of 16°Plato.

This needs to be produced from an all-malt grist, and the malt has an extract of 77%.

To calculate the total amount of extract needed, multiply the volume by the degrees-Plato value, i.e., $800 \times 16 = 12{,}800$ (or 1.28×10^4) hl degrees.

Had the malt been capable of total dissolution (100% extract), we would have needed 1.28×10^4 kilograms.

However, the malt is 77% extract.

Therefore the weight of malt needed is $(1.28 \times 10^4)/ 0.77 = 1.66 \times 10^4$ kg.

The addition of cool malt to hot water leads to a net temperature between the two. The brewer needs to be able to calculate the net temperature that will result from mashing-in, thereby allowing computation of the required "strike" temperature of the water. Similar considerations apply to the addition of adjunct mash to the main mash.

The relevant equation is

$$T_t = \frac{(S \times T_g) + (R \times T_w) + X}{R + S}$$

Where T_t = target starting mash temperature
  S = specific heat of the malt grist (see Table 9)
  T_g = temperature of the grist
  R = ratio of water: grist (hectoliters/100 kg)
  T_w = temperature of the water
  X = slaking heat correction (see Table 9)

Example

We aim to mash in, at 63°C and a water-grist ratio of 3:0, a malt that contains 3% moisture and is at 20°C. Therefore, the target temperature of the striking water (T_w) is found from

$$63 = \frac{(0.40 \times 20) + (3.0 \times T_w)}{3.0 + 0.40} + 2.1$$

or

$$63 = \frac{8 + 3.0\, T_w}{3.4} + 2.1$$

Rearranging

$$T_w = \frac{[(63 - 2.1) \times 3.4] - 8}{3.0}$$

Therefore the temperature of the striking water needs to be 66.4°C.

Further Reading

McCabe, J. T., ed. (1999) The Practical Brewer. Master Brewers Association of the Americas, Wauwatosa, WI.

8. Water

Much more water is used in brewing than ends up in the beer. An efficient brewery uses five times more water than finishes as beer, whereas for a badly run operation the factor may be 20 times. (Don't overlook the huge demand for water in the malting operation, where large amounts of water are needed for steeping, with consequent high-effluent discharges, especially from the first steep.)

Because water represents at least 90% of the composition of most beers, it will clearly have a major direct impact on the product, in terms particularly of flavor and clarity. The water, however, exerts its influence much earlier in the process, through the impact of the salts it contains on enzymic and chemical processes, through the impact on pH, etc.

Water in breweries comes either from wells owned by the brewer (cf., the famous water of Burton-on-Trent in England or Pilsen in the Czech Republic) or from municipal supplies.

Potable Supplies

Water used in the production of beer must adhere to pertaining health standards.

Some brewers differentiate their water supplies, applying greater stringency to the production and cleaning water (i.e., any water that will contact the product either directly in the process stream, *production water,* or in tank- and pipe-cleaning detergents, *cleaning water*) than to the *service water* (used for raising steam, cooling, cleaning of floors, etc.).

The World Health Organization issued guidelines for potable water in 1984, whereas in the United States the NPDWR (National Primary Drinking Water Regulations) standards for potable water date to 1985 (Table 10).

Water Hardness

Hard water is water that does not lather easily with soap, whereas soft water does. The difference is due to the presence of many more calcium and magnesium salts in hard water.

Permanent Hardness

Permanent hardness is caused by the sulfates and chlorides of calcium and magnesium. When water is boiled, they do not change, save that they get more concentrated as water evaporates. Such water is found in regions of high gypsum content.

Table 10. Extract from the National Primary Drinking Water Regulations[a]

Component	Maximum Contaminant Level Goal	Maximum Contaminant Level (mg/L unless stated)	Potential Health Effects	Sources of Contaminant
Cryptosporidium or Giardia	0	99–99.9% removal/inactivation	Diarrhea; vomiting; cramps	Fecal waste
Legionella	0	Deemed to be controlled if Giardia is defeated	Legionnaire's Disease	Multiplies in water-heating systems
Coliforms (including Escherichia coli)	0	No more than 5% samples positive within a month	Indicator of presence of other potentially harmful bacteria	Coliforms naturally present in the environment; E. coli comes from fecal waste
Turbidity	n/a	<1 nephelometric turbidity unit	General indicator of contamination; including by microbes	Soil runoff
Bromate	0	0.01	Risk of cancer	Byproduct of disinfection
Chlorine	4	4	Eye/nose irritation; stomach discomfort	Additive to control microbes
Chlorine dioxide	0.8	0.8	Anemia; nervous system effects	Additive to control microbes
Haloacetic acids (e.g., trichloracetic)		0.06	Risk of cancer	Byproduct of disinfection
Trihalomethanes		0.08	Liver, kidney, or central nervous system ills; risk of cancer	Byproduct of disinfection
Arsenic		0.05	Skin damage; circulation problems; risk of cancer	Erosion of natural deposits; runoff from glass and electronics production wastes
Asbestos	7 million fibers per liter	7 million fibers per liter	Benign intestinal polyps	Decay of asbestos cement in water mains; erosion of natural deposits

Copper	1.3	1.3	Gastrointestinal distress; liver or kidney damage	Corrosion of household plumbing systems; erosion of natural deposits

Contaminant	MCLG	MCL	Health Effects	Sources
Copper	1.3	1.3	Gastrointestinal distress; liver or kidney damage	Corrosion of household plumbing systems; erosion of natural deposits
Fluoride	4	4	Bone disease	Additive to promote strong teeth; erosion of natural deposits
Lead		0.015	Kidney problems; high blood pressure	Corrosion of household plumbing systems; erosion of natural deposits
Nitrate	10	10	Blue Baby Syndrome	Runoff from fertilizer use; leaching from septic tanks; sewage; erosion of natural deposits
Nitrite	1	1	Blue Baby Syndrome	Runoff from fertilizer use; leaching from septic tanks; sewage; erosion of natural deposits
Selenium	0.05	0.05	Hair or fingernail loss; circulatory problems; numbness in fingers and toes	Discharge from petroleum refineries; erosion of natural deposits; discharge from mines
Benzene	0	0.005	Anemia; decrease in blood platelets; risk of cancer	Discharge from factories; leaching from gas storage tanks and landfills
Carbon tetra-chloride	0	0.005	Liver problems; risk of cancer	Discharge from chemical plants and other industrial activities
Dinoseb	0.007	0.007	Reproductive difficulties	Runoff from herbicide use
Dioxin	0	0.00000003	Reproductive difficulties; risk of cancer	Emissions from waste incineration and other combustion; discharge from chemical factories
Alpha particles	0 (as of 12/8/03)	15 picoCuries per liter	Risk of cancer	Erosion of natural deposits
Beta particles and photon emitters	0 (as of 12/8/03)	4 millirems per year	Risk of cancer	Decay of natural and man-made deposits

[a] The full table can be found at http://www.epa.gov/safewater/mcl.html and includes items not listed above, a total of 63 other line items, the majority a range of industrial and herbicidal chemicals.

Temporary Hardness

Temporary hardness is caused by bicarbonates of calcium and magnesium. If such solutions are boiled, then the bicarbonate decomposes, releasing carbon dioxide and precipitating carbonates. Such water is found in limestone, chalk, or dolomite regions.

Hardness can be measured by titration with the chelating agent EDTA.

Production Water

Water is used for
- mashing-in
- sparging
- slurrying or dissolving additions (yeast, foam stabilizer, filter aids, etc.)
- diluting high gravity streams to "sales gravity"

The ionic composition of the water in four brewing centers is given in Table 11.

The water in Burton is clearly very hard, with both permanent and temporary hardness. By contrast the water in Pilsen is extremely soft. The hardness of water (directly and through its influence on pH) influences factors such as the stability of enzymes, the extractability of grist and hop components, the isomerization of hop resins, and the flocculation of yeast.

The water composition can be adjusted, either by adding or removing ions (see later). Thus, calcium levels may be increased in order to promote the precipitation of oxalic acid and to promote amylase action. The alkalinity of water used for sparging may be reduced to less than 50 ppm in order to limit the extraction of tannins.

Furthermore, water may need to be of different standards for different purposes. The microbiological status of water used for slurrying yeast or for use downstream generally is important. Water used for diluting high-gravity streams must be of low oxygen content, and its ionic composition is critical.

Cleaning Water

Cleaning water must be of potable standard and devoid of any off-flavor. Water used for dissolving detergents should be soft and that for disinfectants also sterile. Any rinse water used downstream should be of sound microbiological quality.

Service Water

Only water used for direct injection of steam into a process component need adhere to potable standards. Otherwise, it should adhere to the recommendations of the plant supplier. Water should be adjusted in composition if there is any risk of corrosion to the

Table 11. Ionic composition (mg per liter) of water

Component	Burton	Pilsen	Dublin	Munich
Calcium	352	7	119	80
Magnesium	24	8	4	19
Sulfate	820	6	54	6
Chloride	16	5	19	1
Bicarbonate	320	37	319	333

plant, e.g., a pasteurizer. Good plant maintenance is essential to avoid leaks of steam or cooling streams into the product.

The Role of Key Individual Ions

It is important to appreciate that ions are contributed by other raw materials as well as by water. Control of the ionic composition of beer and the process stream demands reliable measurement of the ionic composition of wort and beer.

There is rather a lot of speculation and unsupported data pertaining to the influence of certain ions. Although they almost certainly play a role, it is by no means certain that manipulating their levels will serve a useful purpose.

Calcium

Calcium promotes α-amylase activity. It is also very important in the precipitation of phosphate and, therefore, reduction of pH:

$$3Ca^{2+} + 2HPO_4^{2-} \rightarrow Ca_3(PO_4)_2 + 2H^+$$

The pH in the brew house is a key determinant of extraction in mashing, hop α-acid utilization, etc.

Calcium is also important for its role in precipitating oxalic acid (from malt). Otherwise, oxalate will precipitate in beer and cause haze and gushing. One recommendation is a 4.5-fold excess of calcium over oxalic acid.

Calcium promotes yeast flocculation and influences precipitation reactions, e.g., in cold conditioning.

Iron

Iron can be detected directly as a metallic flavor when present in concentrations in excess of approximately 0.5 ppm. It is very active as a pro-oxidant, causing hazes and staling at much lower concentrations.

Copper

Copper is also a pro-oxidant but is also capable of binding sulfur compounds such as hydrogen sulfide and lessening their impact.

Zinc

Zinc is a key cofactor in yeast alcohol dehydrogenase. Therefore, it is widely used as a fermentation promoter. It also impacts flocculation and stabilizes foam (promotes lacing).

Bicarbonate

Bicarbonate promotes alkalinity in mash, i.e., higher and perhaps less favorable pH, and extraction of tannins on lautering.

Residual alkalinity = bicarbonate − [(calcium/3.5) + (magnesium/7.0)]

Bicarbonate increases pH, whereas calcium and magnesium lower it (see above). The higher the residual alkalinity, the greater the total alkalinity relative to hardness and therefore the higher the pH.

Sulfate
Sulfate is believed to promote palate dryness and astringency. It is a precursor of sulfur dioxide production by yeast.

Chloride
Chloride is believed to promote palate-fullness and body.

Nitrate and Nitrite
Maximum limits for nitrates/nitrites are set because of their potential carcinogenic effect (nitrate by reduction to nitrite). They are also implicated in the production of nitrosamines.

Phosphate
Phosphate affects pH (see earlier).

Silicate
Silicate from malt can cause haze but is also thought to be healthful (in bone formation and preservation).

Water Treatment

Boiling reduces temporary hardness. Addition of lime (calcium hydroxide) has a similar effect, precipitating carbonates. Acid may be added to reduce alkalinity and temporary hardness. If water is rich in calcium bicarbonate, addition of sulfuric acid will convert temporary into permanent hardness.

Calcium sulfate or calcium chloride may be added to enhance the calcium level, lower pH, or adjust the chloride-sulfate ratio (supposedly a key palate determinant in beer).

Sodium bicarbonate converts permanent hardness into temporary hardness.

Demineralization can be effected by ion exchange resins.

Adsorption (e.g., on activated carbon, charcoal) is used by many brewers to decolorize water and remove undesirable taints.

Precipitation reactions may be employed to remove certain ions: e.g., iron can be eliminated by aeration, coagulation, and subsequent filtration.

Membranes may be used to adjust the ion composition of water (including the elimination of nitrate) and also to remove microorganisms. Techniques include reverse osmosis, electrodialysis, and ultrafiltration.

Deaeration is important for the preparation of dilution water for high-gravity brewing. There are some brewers who also believe this is important for mashing-in and sparge liquor in order to avoid the opportunity for oxidation at this stage. Techniques include driving off oxygen by heat or vacuum, purging with CO_2, and vapor stripping.

Sterilization of water (e.g., for cleaning or for addition to product downstream) may be achieved by chlorination, ozonation, UV treatment, or silver treatment.

Further Reading
Moll, M. (1994) Beers and Coolers. Intercept Ltd., Andover, Hampshire, UK.

9. Hops

A Solitary Outlet

The hop is remarkable among agricultural crops in that essentially its sole outlet is for brewing.

Although hopping accounts for much less than 1% of the price of a pint of beer, it has a disproportionate effect on product quality. Much attention has been lavished on the hop and its chemistry.

Hops are grown in all temperate regions of the world. The major producer is Germany, followed by the United States in three major sites: Washington, Oregon, and Idaho. Hops are grown in the southern hemisphere, with significant crops in Australia and, to a lesser extent, New Zealand and South Africa.

The Hop

There are two separate species of hops: *Humulus lupulus* and *Humulus japonicus*. *H. japonicus* contains no resin and is merely ornamental.

H. lupulus is rich in resins, the source of bitterness, and oils, the source of aroma.

The hop genus (*Humulus*) is within the family Cannabinaceae, and a close relative of the hop is *Cannabis sativa*, Indian hemp, better known as marijuana or hashish.

Cultivation

Hops are hardy, climbing, herbaceous perennial plants. They are grown in yards using characteristic string frameworks to support them (Fig. 53). Their rootstock remains in the ground year after year and is spaced in an appropriate fashion for effective horticultural procedures (for example, spraying by tractors passing between rows).

Dwarf varieties retain the bittering and aroma potential of "traditional" hops but grow to a lower height (6–8 feet as opposed to 12–16 feet). As a result, they are much easier to harvest and there is less waste of pesticide during spraying. Dwarf hop yards are also much cheaper to establish.

Hops are susceptible to a wide range of diseases and pests. The most serious diseases are Verticillium wilt, downy mildew, and mold. The most serious pest is the damson-hop aphid (hop fly), which in the fall lays eggs in the bark of the damson (or sloe or plum). In the spring, the flies hatch and migrate to the hop plants.

Varieties differ in their susceptibility to infestation and have been progressively selected on this basis. Nonetheless, it is frequently necessary to apply pesticides, which are always stringently evaluated for their influence on hop quality, for any effect they may have on the brewing process, and, of course, for their safety.

84 / Chapter 9

Fig. 53. Trellis work in a hop garden. (Courtesy of Yakima Chief)

Downy mildew is caused by a fungus, *Pseudoperonospora humuli*. The disease is rapidly transmitted among parts of plants and separate plants. Infection may prevent the development of cones and, where cones do develop, yields of the α-acid precursors of bitter compounds are reduced. Infected stock must be burnt—and the grower must realize that mycelia may persist in rootstock and demand attention in subsequent years.

Hop mold is caused by the fungus *Sphaerotheca macularis*, revealing itself as red patches on leaves and cones. One effective treatment involves spraying with sulfur.

Wilt is caused by *Verticillium albo-atrum*. Again, it can spread rapidly and requires urgent attention if it is to be contained. Infection demands the burning of plants and infected rootstock.

The Active Ingredients of Hops

The components of the hop required by the brewer—the resins and the oils—are located in the cones of female plants (Fig. 54). More particularly, they are found in the lupulin glands that are alongside the seeds at the base of the bracteoles.

In the United Kingdom, male plants are planted alongside female plants, leading to fertilization and seed levels of up to 25%. Hops are perennial, however, and can be propagated from cuttings. In the rest of Europe and the United States (with some exceptions in Oregon), there is no planting of male hops, and the hops supplied for brewing are seedless. On a weight-by-weight basis, the content of resin and oil is greater in seedless hops, but horticultural yield is lower. It is believed by some that seedless hops make for easier downstream processing of beer.

The Hop-Grower's Year

A typical hop-grower's year in the hop-growing district of Yakima in the state of Washington begins in March, with some shallow plowing to lower the weed count and to

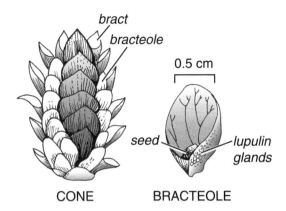

Fig. 54. Hop cone structure.

mulch into the ground residual leaves and vines from the previous crop and incorporate some fertilizer.

In the following month, the wirework will be established on 3-m wooden poles at a spacing of 2 m × 2 m. New shoots from the rootstock are then trained onto the strings.

In June, the fields are plowed to control weeds, and spraying will be done in July and August to control pests.

Harvest will commence in mid-August and last for a month. Traditionally, this was a most labor-intensive operation, but now machine-picking is universally employed.

Hops are dried according to principles similar to those for barley and malt. Using temperature regimes between 55 and 65°C, the moisture content will be reduced from 75 to 9%.

In the United States, hops are packed into bales measuring (in inches) 20 × 30 × 57 and weighing 200 pounds (91 kg or 1.8 Zentners; 1 Zentner = 50 kg of hops).

Hop Analysis

The analysis of hops remains somewhat more primitive than that of other brewery ingredients. Many of the assessment criteria for hop quality depend on non-instrumental judgment by experts.

First, the sample of hops is inspected visually for signs of deterioration, infestation, or weathering. Then a sample is rubbed between the palms of the hands before the contents are sniffed; the assessor is looking for any smells associated with deterioration and is determining whether the "nose" is consistent with that which is expected from the variety in question.

The prime quantitative parameter upon which hop transactions are made is the content of α-acids, namely, the precursors of the bitter compounds. In the United States, α-acids are measured spectrophotometrically: the extent to which an extract of hops absorbs ultraviolet light of three wavelengths indicates the amount of resins present (Fig. 55). In the United Kingdom, α-acids are measured by the lead conductance test (Fig. 56).

Types of Hops

All hops are capable of providing both bitterness and aroma. However, they are frequently classified into aroma hops, bittering hops, and dual-purpose hops.

Some hops, such as the Czech variety Saaz, have a relatively high ratio of oil to resin, and the character of the oil component is particularly prized. Such varieties command higher prices and are known as *aroma varieties*. They are seldom used as the sole source of bitterness and aroma in a beer.

A cheaper, higher-α-acid hop (a *bittering variety*) is used to provide the bulk of the bitterness, with the prized aroma variety added late in the boil for the contribution of its own unique blend of oils. Those brewers requiring hops solely as a source of bitterness

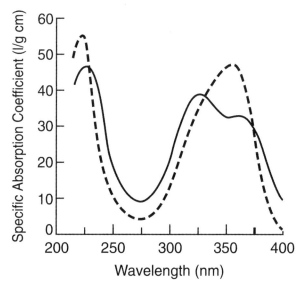

Fig. 55. Spectrophotometric analysis of hop acids. Dotted line = β-acid (alkaline); solid line = α-acid (alkaline). The equations for acid content are α-acid (%) = $d \times (-51.56A_{355} + 73.79A_{325} - 19.07A_{275})$ and β-acid (%) = $d \times (55.57A_{355} - 47.59A_{325} + 5.10A_{275})$, where d is the dilution factor.

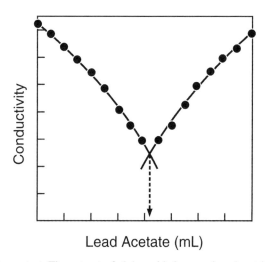

Fig. 56. The lead conductance test. The extract of alpha acids in organic solvent is titrated with lead acetate solution as the conductance is measured. As the negatively charged hop acid is neutralized with lead ions, the conductance decreases. Once all the acid is converted to the lead salt, the conductance starts to rise again. Thus, the alpha acid content is proportional to the volume of lead acetate solution used.

may well opt for a cheaper variety, ensuring its use early in the kettle boil so that the provision of bitterness is maximized and unwanted aroma is driven off.

Dual-purpose hops tend to have intermediate levels of resins together with an appealing aroma delivery.

Hop Chemistry

Hops contain a range of chemical species, including cellulose and lignin as structural components, proteins, lipids and waxes, and tannins. We need consider only two constituents: the resins and the essential oils.

Resins. There are several components in the resin fraction of the hop (Figs. 57 and 58). Most brewers consider only one type of component as being significant: the α-acids. These molecules, also known as the humulones, can account for as little as 2% of the dry weight of the hop or as much as 15%.

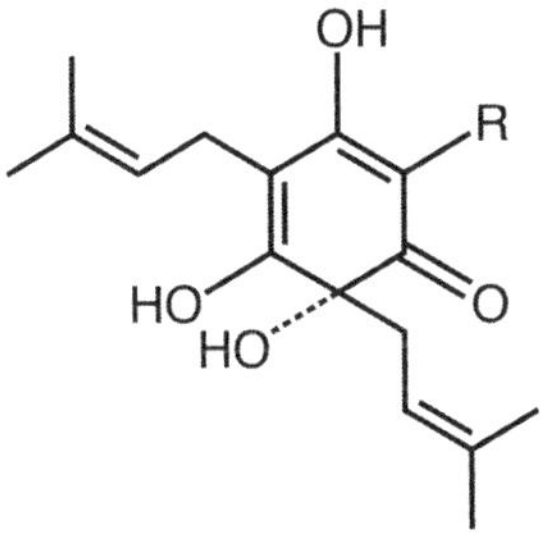

Name	Side Chain (R)
Humulone	—CO·CH$_2$·CH(CH$_3$)$_2$ isovaleryl
Cohumulone	—CO·CH(CH$_3$)$_2$ isobutyryl
Adhumulone	—CO·CH(CH$_3$)·CH$_2$·CH$_3$ 2-methylbutyryl

Fig. 57. Alpha acids.

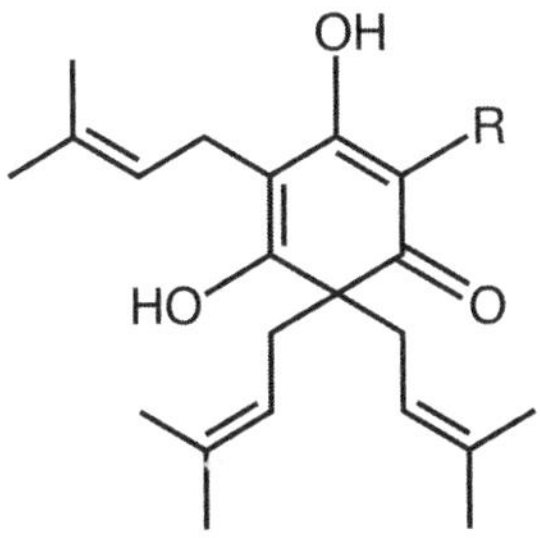

Name	Side Chain (R)
Lupulone	—CO·CH$_2$·CH(CH$_3$)$_2$ isovaleryl
Colupulone	—CO·CH(CH$_3$)$_2$ isobutyryl
Adlupulone	—CO·CH(CH$_3$)·CH$_2$·CH$_3$ 2-methylbutyryl

Fig. 58. Beta acids.

A "high-α-acid" variety is a richer source of bitterness. Less of it will need to be used to impart a given bitterness level to beer, but of course there will be a proportionately lower contribution of the essential oils, i.e., less aroma potential.

Conversely a low-α-acid hop needs to be used in larger proportions to afford a desired bitterness, which leads to greater potential aroma delivery. This is at the risk of introducing other undesirable materials, such as the tannoids that promote haziness in beer.

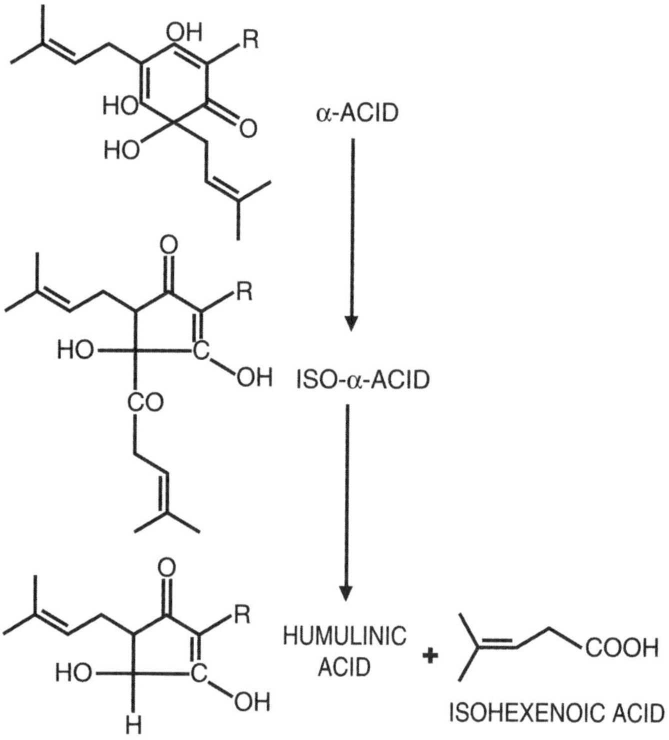

Fig. 59. Degradation of alpha acids.

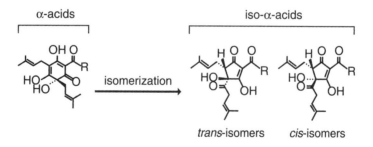

R = CH$_2$CH(CH$_3$)$_2$	humulone	isohumulone
R = CH(CH$_3$)$_2$	cohumulone	isocohumulone
R = CH(CH$_3$) CH$_2$CH$_3$	adhumulone	isoadhumulone

Fig. 60. Isomerization of alpha acids.

α-Acids break down during the storage of hops, leading to a loss of bittering potential and the release of short-chain fatty acids that afford "cheesy" character (Fig. 59). Deterioration is promoted by excessive moisture, warmth, and air.

There are three variants of the α-acids (cohumulone, humulone and adhumulone), differing in the structure of the R side chain. Received wisdom contends that "better" hops have a relatively low proportion of cohumulone.

When wort is boiled in the kettle, the α-acids are rearranged to form iso-α-acids in a process referred to as "isomerization" (Fig. 60). The products are much more soluble and are more bitter. At the end of boiling, any unisomerized α-acid is lost with the spent hop material and the iso-α-acids remain. The process is inefficient, with perhaps no more than 50% of the α-acids being converted in the boil and less than 25% of the original bittering potential surviving into the beer.

Each iso-α-acid exists in two forms, *cis* and *trans*, which differ in the orientation of the side chains. The six iso-α-acids differ in the quality and intensity of their bitterness (see later).

Essential oils. The oil component of hops can amount to up to 3% of the weight of a hop. Seedless hops tend to contain more essential oil. The oils are produced in the hop late in ripening, after the majority of the resin has been laid down, which highlights the need for harvesting of the hops at the appropriate time.

The oil is a complex mixture of at least 300 compounds (Fig. 61). Nobody has established a firm relationship between the chemical composition of the essential oils and the unique aroma characteristics that they deliver.

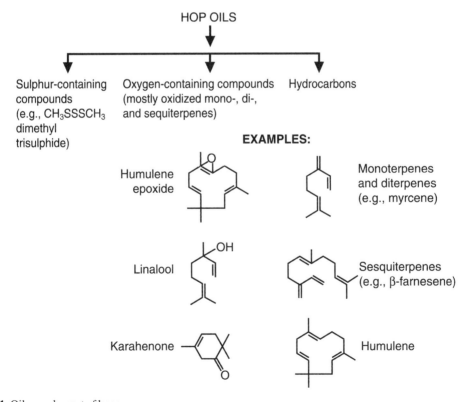

Fig. 61. Oil complement of hops.

It is most likely that "late hop character" (i.e., that aroma associated with lagers from Europe, which is introduced by adding a proportion of the hops late in the boil) is due to the synergistic action of several oil components, perhaps modified by the action of yeast in the ensuing fermentation.

"Dry hop character" (a feature associated with traditional English cask ales, afforded by adding a handful of whole hop cones to the finished beer) is no less complicated.

Generally it is felt that myrcene, the major hydrocarbon component, is an undesirable feature of the oil, whereas linalool and geraniol, present in far lower concentrations, offer attractive aroma notes.

To a greater or lesser extent, the individual essential oil components are lost from wort during boiling. The delivery of a given hop character, then, depends on the skill of the brewer in adding the hops at exactly the right time to ensure survival of the right mix of oils that imparts a given character to the product.

Hop Preparations

The use of whole-cone hops (Fig. 62) is comparatively rare nowadays.

The most common procedure for hopping is to add hops that have been hammer-milled and then compressed into pellets (Fig. 63). In this form they are more stable, more efficiently utilized, and they do not present the brewer with the problem of separating out the vegetative parts of the hop plant.

Nevertheless, because of the inefficient utilization of the α-acids during wort boiling, even from pellets, and as a result of vagaries in the introduction of defined hoppy aromas into beers, a wide selection of hop preparations has reached the marketplace.

Extracts are mostly based on the prior extraction of hops with liquid or supercritical carbon dioxide (Fig. 64). The resultant extracts can be fractionated into hop resin- and oil-rich fractions; the resin portion is available as a source of bitterness for addition in place of whole hops or pellets to the kettle, and the oil part provides an opportunity for

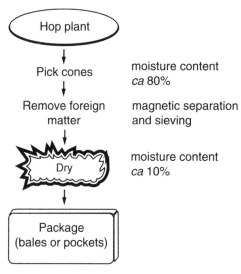

Fig. 62. Packaging of whole hops.

controlled addition of hop character, either by dosing late in the boil for a late hop character or into the finished beer for a dry hop note (Fig. 65).

It is possible to carry out the isomerization of the α-acids in the liquid CO_2 extracts by chemical means or by the use of light (Fig. 66). Therefore, it is possible using the resultant "pre-isomerized extracts" to add bitterness directly to the finished beer, which allows far better utilization of the bitter compounds, because the extent of isomerization of α-acids is greater and because bitter substances are no longer lost by sticking onto yeast cells and trub.

Recent years have been marked by an enormous increase in the use of such pre-isomerized extracts after they have been modified by reduction (Fig. 67). One of the side-chains on the iso-α-acids is susceptible to cleavage by light; it then reacts with traces of

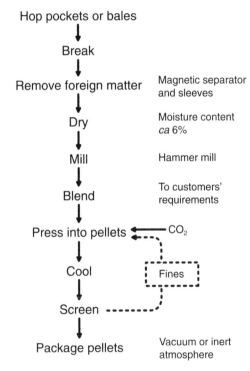

Fig. 63. Production of pellets.

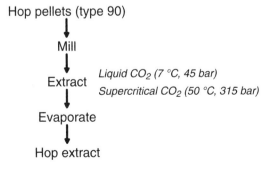

Fig. 64. Production of hop extracts.

sulfidic materials in beer to produce 2-methyl-3-butene-1-thiol (MBT), which imparts an intensely unpleasant skunky character to beer. If the side-chain is reduced, it no longer produces MBT.

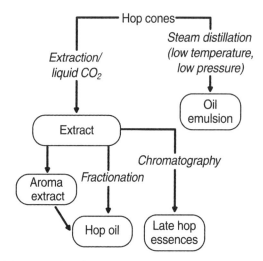

Fig. 65. Production of oil-rich extracts from hops.

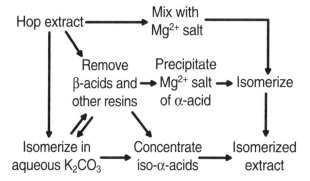

Fig. 66. Isomerization of extracted resins.

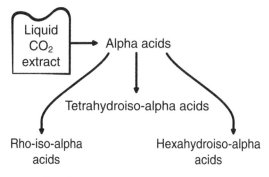

Fig. 67. Reduction of resin extracts from hops.

Late hop aroma can be introduced through the use of extracts, too. The essential oils can be split into two fractions, one spicy and the other floral. By adding them to bright beer in different proportions, it is possible to impart different late hop characters.

Further Reading

Neve, R. A. (1991) Hops. Chapman and Hall, London

10. Wort Boiling, Clarification, and Cooling; Sugars

The boiling of wort serves various functions:
- inactivation of residual enzymes—"fixing" of wort composition
- isomerization of α-acids from hops
- sterilization
- removal of unwanted volatile materials
- precipitation of proteins/polyphenol complexes (as "hot break" or "trub")
- concentration

There is also a degree of color formation in boiling through Maillard reactions. Insoluble materials produced in the boil are removed during clarification prior to cooling to the temperature required for fermentation. It is customary at the cooling stage to introduce either air or oxygen to promote yeast activity. During cooling, more insoluble material falls from solution ("cold break").

Wort Boiling

The extent of wort boiling is normally described in terms of percent evaporation. Water is usually boiled off at a rate of about 4% per hour, and the duration of boiling is likely to be 1–2 hr. Factors determining the length of boiling include, for example, the extent to which the brewer wishes to eliminate dimethyl sulfide from the beer (see later). At 100°C in wort of pH 5.2, approximately 1% of the total α-acid is isomerized every minute.

Brew Kettles

Sometimes brew kettles are referred to as "coppers," reflecting the original metal from which they were fabricated. These days, they are usually made from stainless steel but sometimes contain some "sacrificial copper" to deal with sulfur compounds.

Kettles comprise some form of heating device, cleaning system, an exhaust for vapors, a gauge, and a manhole cover. Some brewers insist that the manhole cover should remain open during boiling, others that it should be closed to prevent air pick-up.

Boiling is the most energy intensive operation in the brewery. Various systems have been introduced in an effort to reduce energy demand, including constant wort boiling, vapor recompression systems, low-pressure boiling, high-pressure boiling, etc. It is essential that any such system have no impact on the quality of the product. The pressure to effect the shortest practical boil is not only for reasons of energy conservation, but also to facilitate turnover in the brew house.

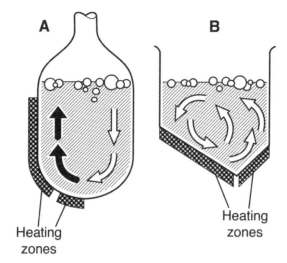

Fig. 68. Kettles with heating jackets.

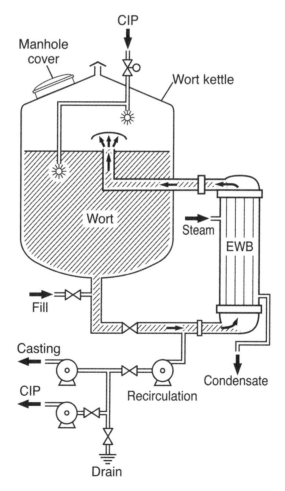

Fig. 69. Kettle with a calandria. EWB = external wort boiler.

Kettles obviously become soiled on successive brews, and there must be regular cleaning (perhaps every third brew) to refresh heating surfaces and remove caramelized soil.

There are many shapes for kettles (e.g., Figs. 68 and 69). The essential criterion, regardless of shape, is the extent to which a "rolling boil" can be achieved. It is only when the boil is vigorous that there will be efficient steam volatilization of off-flavors, effective precipitation of proteins and polyphenols, etc.

Systems can have either internal or external steam heating; the devices are sometimes called *calandria*, being tube and shell heat-exchanger systems (Fig. 69). Wort movement is convective. An external device can service more than a single kettle. Kettles using jacket heating incorporate asymmetric positioning of the heating surfaces as this, too, allows convective currents to be established in the wort (see Fig. 68). There is limited use of direct firing of kettles (i.e., direct flame heating of a vessel).

Additions at the Wort Boiling Stage

Certain fining materials (e.g., carrageenan; Fig 70) may be added to promote protein precipitation. Extra calcium may be added to promote pH drop.

Liquid sugar adjuncts may be added. These are called "wort extenders"; they increase the extract from a brew house without investment in extra mashing vessels and wort separation devices. They do require installation of storage tanks.

Sugars must be added carefully such that they are mixed in evenly and without caramelization. Addition must be slow to prevent a sudden drop in temperature.

Most sugars are derived from either corn or sugar cane. In the latter case, the principle sugar is either sucrose or fructose plus glucose if the product has been "inverted."

There are many different corn sugar products, differing in their degree of hydrolysis and therefore fermentability (Table 12). Through the controlled use of acid, but increasingly of starch-degrading enzymes, the supplier can produce preparations with a

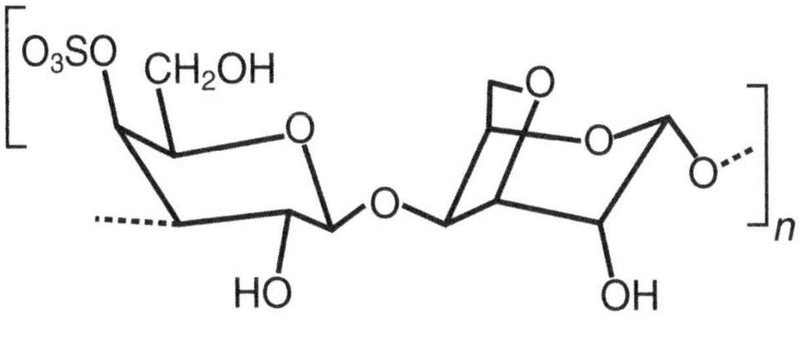

Fig. 70. Carrageenan.

Table 12. A selection of brewing sugars and syrups

	% Glucose	% Maltose	% Maltotriose	% Dextrins	Other
Dextrose	100	–	–	–	–
Corn syrup	45	38	3	14	–
High maltose	10	60	–	30	–
Maltodextrin	~0	1.5	3.5	95	–
Sucrose	0	0	0	0	100% sucrose
Invert sugar	50	0	0	0	50% fructose

full range of fermentabilities, depending on the needs of the brewer, from 100% glucose through to high-dextrin/low fermentability.

Wort Clarification

After boiling, wort is transferred ("knockout") to a clarification device. The system employed for removing insoluble material after boiling depends on the way in which the hopping was carried out.

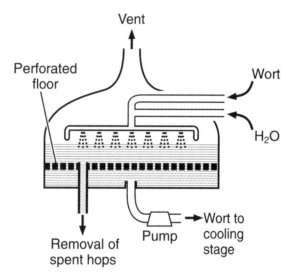

Fig. 71. Hop back.

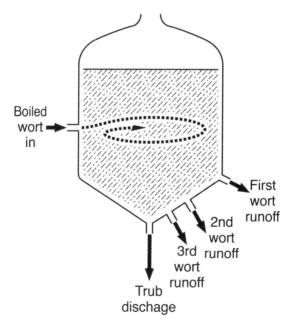

Fig. 72. Principle of a whirlpool separator.

If whole hop cones are used, primary clarification is through a hop jack (hop back, Fig. 71). This is analogous to a lauter tun, but in this case the bed of residual hops constitutes the filter medium. Sparging is used to ensure optimum wort recovery.

If hop pellets or extracts are used, the device of choice is the whirlpool, although previously a "cool ship" might have been used. A cool ship comprises a vast rectangular vessel at the top of the brewery, which holds less than a foot of wort left for up to half a day to lose its heat. This process is very unhygienic and hence very rare these days. Centrifuges or hydrocyclones could also be used to remove particles.

The whirlpool is a cylindrical vessel, typically about 5 m in diameter, into which hot wort is transferred tangentially through an opening 0.5- to 1-m above the base (Fig. 72). The wort is set into a rotational flux, which forces trub to a pile in the middle of the vessel. After a period of up to 1 hr, the bright wort is run off from the vessel from pipes in the base, without disturbing the trub. The vessels may be insulated in order to conserve a high temperature and prevent re-dissolution of the hot break.

One major brewer employs wort stripping, in which hot wort in a thin film has air passed through it. This eliminates unwanted volatiles that survive (or are developed in) the boiling and clarification stages.

Wort Cooling

Almost all cooling systems today are of the stainless steel plate heat exchanger type (Fig. 73). These systems are sometimes called "paraflows." Such devices use either water alone (if the temperature of the worts needs to attain 15°C) or have a second stage (coolant—refrigerated water, brine, or propylene glycol) if the temperature needs to reach as low as 6°C. Heat is transferred from the wort to the coolant. With water as coolant, the conserved heat is used in steam raising. The wort flow should be turbulent. Effective maintenance of the system is essential if leakage of coolant into the wort is to be avoided. The device will need cleaning (e.g., every eight brews) to ensure that heat transfer efficiency is high.

At this stage, it is likely that more material will precipitate from solution ("cold break"). Brewers are divided on whether they feel this to be good or bad for fermentation and beer quality. The presence of this break certainly accelerates fermentation, and

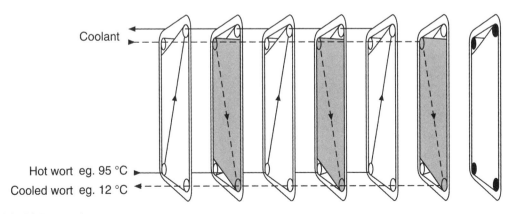

Fig. 73. Heat exchanger.

therefore it will directly influence yeast metabolism. As in so much of brewing, the aim should be consistency: either consistently "bright worts" or ones containing a relatively consistent level of trub.

Wort is usually either aerated or oxygenated at this stage, depending on the amount of oxygen needed by the yeast (see later). Injection can either be on the hot side of the paraflow or the cold side. The former presents less of a microbiological risk, and there may be a better quality of cold break subsequently formed. Conversely, gases are more soluble as the temperature falls, and there is less risk of oxidative damage to wort components if the gas is put into the cold stream and is soon made available to the yeast.

Further Reading

Andrews, J., and Axcell, B. C. (2003) Wort boiling–Evaporating the myths of the past. Tech. Q. Master Brew. Assoc. Am. 40:249-254.

11. Yeast

The brewing yeast genus is *Saccharomyces* ("sugar fungus").

There are many separate strains of brewing yeast, each of which is distinguishable in some way:
- *phenotypically*—e.g., in the extent to which it will ferment different sugars, in the amount of oxygen it needs to prompt its growth, in the amounts of its metabolic products (i.e., flavor spectrum of the resultant beer), or its behavior in suspension (top vs. bottom fermenting, flocculent or non-flocculent)
- *genotypically*—in terms of its DNA fingerprint

Taxonomists now classify ale strains as *S. cerevisiae* and lager strains as *S. pastorianus*. The latter seems to have evolved through a merging of *S. cerevisiae* and a strain sometimes used by winemakers, *S. bayanus*. However, there are those who still talk in terms of *S. uvarum* or *S. carlsbergensis* as lager yeasts.

The fundamental differentiation between ale and lager strains has been based on the ability or inability to ferment the sugar melibiose. Only lager strains can do this, because they contain the *MEL* gene that codes for an extracellular enzyme α-galactosidase, which splits melibiose into glucose and galactose, both of which can be used by yeast but not when they are joined together (Fig. 74).

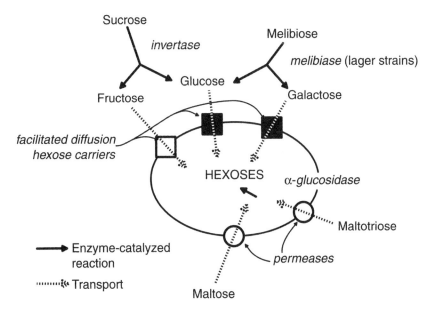

Fig. 74. Access of sugars to yeast cells.

Historically, brewers differentiated ale and lager yeasts functionally on the basis of their migratory tendencies in a fermenter. Ale yeasts moved to the top of the fermentation vessel when fermenting dark worts from well-modified malts at temperatures often about 18–22°C. They were called top-fermenting yeasts. Lagers were fermented with yeasts that dropped to the bottom of fermenters during fermentation at 6–15°C of paler worts from less well-modified malts, and they were termed bottom-fermenting yeasts.

Nowadays, it is frequently difficult to make this differentiation, because beers are widely fermented in similar types of vessels (deep cylindroconical tanks) at similar temperatures and from worts that may be of diverse types and produced from a range of grist materials.

Yeast Structure

A schematic drawing of yeast as seen under an electron microscope is shown in Figure 75. A yeast cell is either spherical or ellipsoidal and may have a diameter of 5–13 μm. It is surrounded by a cell wall that provides protection, determines the shape, and regulates the extent to which the yeast cell will interact with other yeast cells and with other materials (e.g., finings, surfaces). The wall largely comprises polysaccharide (60%), together with protein and lipid. There is mannoprotein (Fig. 76) and a β-glucan with 1-3 and 1-6 links (i.e., different from the glucan in the walls of the starchy endosperm of barley, Fig. 77).

Within the wall is a plasma membrane, principal function of which is to regulate what enters and leaves the cell. In common with most other membranes, it contains lipid and protein. A key component in the lipid fraction is sterol.

The genetic material of the cell is largely contained in the nucleus. The DNA is held in 16 small chromosomes. Laboratory strains of yeast contain just one set of chromosomes

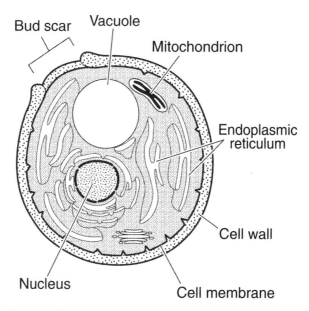

Fig. 75. Cross section of a yeast cell.

and are said to be "haploid." However, brewing strains are "aneuploid": they contain at least three copies of the genome, but in some strains there are more or fewer copies of individual chromosomes.

In common with other aerobic organisms, brewing yeast does develop mitochondria. In aerobically grown cells (i.e., yeast that is respiring), there may be 50 or so mitochondria per cell. However, if yeast is in fermentative mode (as in a brewery fermentation), then the mitochondria are fewer and possess a different, more elongated morphology with less well-developed internal membranes.

Yeast Life Cycle

Yeast reproduces by budding. A single cell may bud many times, and each time a scar is left on the surface of the mother cell (not the progeny). A cell may display 10–40 bud scars, and the number is a useful index of the age of a population of yeast.

A haploid yeast cell presented with optimal environmental conditions buds in approximately 90 min (i.e., this is the doubling time). Yeast has a well-understood life cycle,

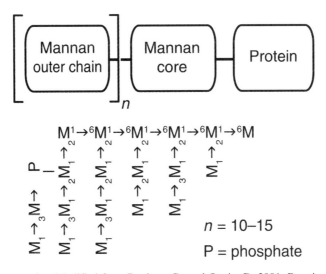

Fig. 76. Yeast mannoprotein. (Modified from Boulton, C., and Quain, D. 2001. Brewing Yeast and Fermentation. Blackwell Science. London)

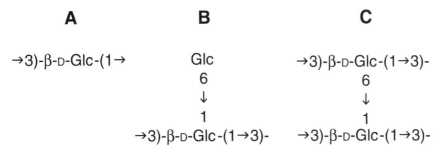

Fig. 77. Structural basis of yeast glucan.

involving two mating types (a and α). It is possible for the scientist to take advantage of these types to breed new strains. Current strain-improvement techniques rely more on cloning technology than "classical breeding."

Strain Selection and Improvement

There are two approaches to isolating new strains for brewing, which we might call "targeted" and "serendipity." In the latter case, a range of strains may be isolated almost at random and tested for their ability to ferment wort and the manner in which they do it (flavor of resultant beer, rate of attenuation, etc.). In the targeted case, a specific need is identified (e.g., ability to produce lower levels of a certain flavor component), and an existing strain is genetically modified in order to meet that requirement.

Classical techniques for improving yeast strains include mutation, hybridization, rare mating, and cytoduction.

More frequently nowadays, the improvement process involves recombinant DNA technology (described in Appendix 1). In brewing, there have been several reports of yeast improvement by recombinant DNA technology, but none has yet been used commercially.

Yeast Nutrition

Yeast requires supplies of
- energy (sugars)
- nitrogen (amino acids and certain peptides)
- lipid (either preformed or produced by the yeast itself from metabolic intermediates, in which case it needs some oxygen that is involved in sterol synthesis and in the desaturation of fatty acids)
- sulfur (sulfate or S-containing amino acids)
- vitamins
- minerals (including zinc)

Carbohydrate Metabolism

Wort contains a selection of carbohydrates, some of which are fermentable. The spectrum of carbohydrates present depends on the grist and on the mashing procedures (see earlier). However, for a typical all-malt wort (produced without the use of exogenous glucoamylase enzyme), the proportion of carbohydrates may approximate to maltose (45%), maltotriose (15%), glucose (10%), sucrose (5%), fructose (2%), and dextrin (23%).

The dextrins are maltotetraose (four glucoses linked α 1-4) and larger such molecules (more glucosyls), and they are unfermentable.

The other sugars ordinarily would be utilized in the sequence sucrose, glucose, fructose, maltose, and lastly maltotriose. There may be some overlap (Fig. 78).

Sucrose is hydrolyzed by an enzyme (invertase) released by the yeast outside the cell, and then the glucose and fructose enter the cell to be metabolized (Fig. 74).

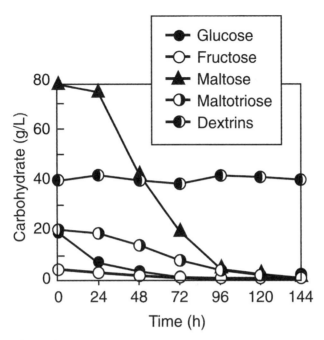

Fig. 78. Consumption of sugars by yeast.

Maltose and maltotriose enter through the agency of specific permeases. Inside the cell, they are broken down into glucose by an α-glucosidase. Glucose represses the maltose and maltotriose permeases.

EMP pathway. The principal route of sugar utilization in the cell is the Embden-Meyerhof-Parnas (EMP) pathway of glycolysis. It is more important to understand the principles of the pathway than its precise details.

The EMP pathway involves the oxidation of sugar, with the capture of energy in the form of ATP. Some ATP is used up initially in the phosphorylation ("activation") of the sugar. There is a net gain of 2 ATP molecules for every molecule of glucose consumed. Oxidation involves a removal of electrons, which are captured by a carrier NAD. In fermenting yeast, the NADH is recycled by reducing the end-product pyruvate to ethanol.

Thus we can divide the process into the following stages:

activation:	sugar + 2ATP → sugar diphosphate + 2ADP
oxidation:	sugar diphosphate + 2NAD + 2 phosphate → 2 diphosphoglycerate + 2NADH
energy capture:	2 diphosphoglycerate + 4 ADP → 2 pyruvate + 4 ATP
NAD regeneration:	2 pyruvate + 2 NADH → 2 ethanol + 2 CO_2 + 2 NAD
Net:	sugar + 2ADP + 2 phosphate → 2 ethanol + 2 CO_2 + 2 ATP

The energy generated in this pathway in the form of ATP is used for the various energy-demanding reactions in the yeast cell, e.g., biosynthesis of cell components, thereby releasing ADP ready for carrying on the EMP pathway. Thus ADP/ATP constitutes the energy vector in the cell.

Nitrogen Metabolism

Yeast derives most of the nitrogen it needs for the synthesis of proteins and nucleic acids from the amino acids in the wort.

A series of permeases is responsible for the sequential uptake of the amino acids. At the start of fermentation, eight of the amino acids (Group A) are taken up rapidly: arginine, aspartic acid, asparagine, glutamic acid, glutamine, lysine, serine, and threonine. Group B, the next to be assimilated, comprises valine, methionine, leucine, isoleucine, and histidine. Group C, which does not start to be consumed until all of Group A is utilized, consists of glycine, phenylalanine, tyrosine, tryptophan, and alanine. Finally there is Group D, consisting only of the amino acid proline, which is not assimilated at all during fermentative growth of yeast.

The amino acids are transaminated to keto acids and held within the yeast until they are required, when they are transaminated back into the amino acid (Fig. 79).

The amino acid spectrum and level in wort (free amino nitrogen) is significant, as it influences yeast metabolism leading to flavor-active products (see later).

Oxygen Requirement of Yeast

In the absence of oxygen, yeast is unable to synthesize the unsaturated fatty acids and sterols that it needs for its membranes. Although these can be introduced into wort, it is usual for oxygen to be introduced into wort (prior to fermentation) in the quantities that the yeast needs for the job in question. It is advisable not to provide too much oxygen because this causes excessive yeast growth—and the more yeast is produced in fermentation, the less alcohol is produced.

Yeasts can be classified into four groups based on their requirement for O_2.

Group 1: requires half air saturation
Group 2: requires air saturation
Group 3: requires O_2 saturation
Group 4: not satisfied even by O_2 saturation

The solubility of oxygen is influenced by wort temperature and composition: the higher the temperature and the more material already dissolved in the wort, then the less

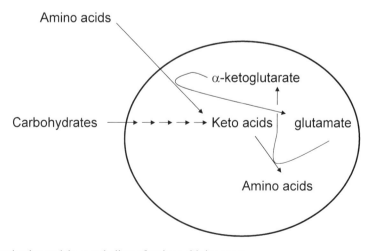

Fig. 79. Transamination and the metabolism of amino acids by yeast.

soluble the oxygen. At 10°Plato and 15°C, though, the amount of oxygen dissolved at air saturation is 7.1 mg per liter; and at oxygen saturation, it is 33.8 mg/liter (remember that air is 80% N_2).

Storage Materials in Yeast

Yeast uses its stored reserves of carbohydrate in order to fuel the early stages of metabolism when it is pitched into wort, e.g., the synthesis of sterols. There are two principal reserves: glycogen and trehalose. Glycogen is similar in structure to the amylopectin fraction of barley starch. Trehalose is a disaccharide comprising two glucoses linked with an α 1-1 bond between them.

The glycogen reserves of yeast build up during fermentation, and it is important that they be conserved in the yeast during storage between fermentations (Fig. 80). When the yeast is pitched into the next fermentation, it does not have all of the necessary "machinery" to deal with the wort—e.g., the permeases needed to transport sugars into the cell. Making this machinery demands energy. As this is not available from the sugars outside the yeast, the yeast must draw on its reserves, namely, glycogen.

Trehalose may function more as a protectant against the stress of starvation. It certainly seems to help the survival of yeast under dehydration conditions employed for the storage and shipping of dried yeast (e.g., in home brewing kits).

Yeast Handling

Pure yeast culture was pioneered by Emil Christian Hansen at Carlsberg in 1883. By a process of dilution, he was able to isolate individual cells and open up the possibility of selecting and growing separate strains for specific purposes.

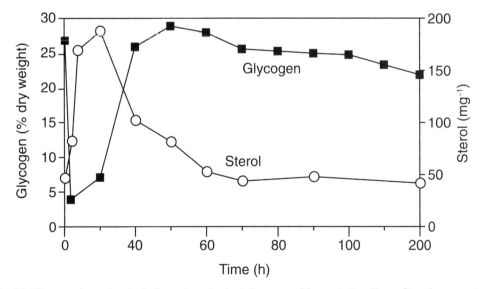

Fig. 80. Glycogen is used up to fuel sterol synthesis at the onset of fermentation. Thereafter, glycogen stocks are replenished.

Although some brewers simply use the yeast grown in one fermentation to "pitch" the following fermentation, and they have done this for many decades, it is much more common for yeast to be repropagated from a pure culture every five to eight generations. There is one company in the United States that uses freshly propagated yeast in every fermentation. (When brewers talk of "generations," they mean successive fermentations; strictly speaking, yeast advances a generation every time it buds, and therefore there are several generations during any individual fermentation batch.)

Storage of Yeast Strains

Having been isolated by micro-manipulation or more commonly by picking off a single colony from a streak plate, the "master culture" of yeast can be stored in various ways, e.g., slope, stab culture, freeze-dried, or under liquid nitrogen. Key criteria are the retention of viability and minimization of genetic variation. The consensus is that deep freeze is best and freeze-drying worst.

Yeast Propagation

Large quantities of yeast are needed to pitch commercial-scale fermentations. They need to be generated by successive scale-up growth from the master culture (Fig. 81).

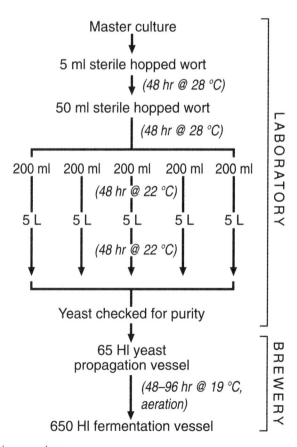

Fig. 81. Yeast propagation overview.

The culture is first checked for its properties to ensure that it is the correct strain. Methods might include giant-colony morphology, flocculation characteristics, or ability to utilize certain nutrients (by using commercial kits or proprietary media).

The propagation cycle is likely to begin with cells from the master culture being introduced to 5 ml of sterile hopped wort and the flask shaken on an orbital incubator in the lab for 48 hr at perhaps 28°C. This is then used to inoculate, successively, 50 ml, 200 ml, and 5 liters of wort with growth for 48 hr at 28, 22, and 22°C, respectively.

The yeast is checked for purity before it leaves the lab stages and is used to inoculate 50 hl of sterile wort in a rigorously cleaned yeast-propagation vessel. There is a facility for plentiful aeration/oxygenation of the yeast, as the requirement here is for yield rather than beer. This phase is likely to take 2–3 days at 19°C and will generate sufficient yeast (8–25 g/liter) to pitch a 500-hl fermenter.

Higher yields are possible if fed-batch culture is used. This is the type of procedure used in the production of baker's yeast. It takes account of the Crabtree effect, in which high concentrations of sugar drive the yeast to use it fermentatively rather than by respiration. When yeast grows by respiration, it captures much more energy from the sugar and therefore produces much more cell material. In fed-batch culture, the amount of sugar made available to the yeast at any stage is low. Together with the high levels of oxygen in a well-aerated system, the yeast respires and grows substantially. The sugar is "dribbled in," and the end result is a far higher yield of biomass (perhaps fourfold more) than is produced when the sugar is provided in a single batch at the start of fermentation.

Acid Washing

The majority of brewing yeasts are resistant to acid (pH 2.0–2.2), treatment with which is very effective in killing bacteria (but not wild yeast). Accordingly, many brewers employ an acid washing of yeast between fermentations. Certain precautions must be taken if the yeast health is not to suffer:
1. food grade acid should be used—phosphoric is favored
2. the yeast slurry and the acid should be chilled to 2–5°C before use, and at no stage should the temperature be allowed to rise
3. continuous stirring is necessary when the acid is being added to the yeast slurry
4. the pH of the slurry should be checked and should be within 0.1 unit of the target
5. washing should be for no longer than 2 hr, and the yeast should be pitched (including the acid) immediately thereafter
6. the yeast must be in good condition (viable)

Yeast Condition

There are two key indices of yeast health: viability and vitality. Both should be high if a successful fermentation is to be achieved.

Viability

Viability is a measure of whether a yeast culture is alive or dead. Whereas microscopic inspection of a yeast sample is useful as a gross indicator of that culture (e.g., presence of substantial infection), quantitative evaluation of viability needs a staining test. The most common is the use of methylene blue: viable yeast decolorizes it, dead cells do not. Recently, some scientists have favored methylene violet, which supposedly registers fewer "false positives." A more accurate procedure is the use of slide culture, in which a diluted suspension of yeast is applied to a microscope slide covered with a layer of nutrient. After 18 hr of incubation, cells can be distinguished as either giving micro-colonies (viable) or not (dead).

Vitality

Although a yeast cell may be living, it may not necessarily be healthy. Vitality is a measure of how healthy a yeast cell is. Many techniques have been advanced as an index of vitality, but none has been accepted as definitive. The techniques include sterol measurement, assessment of acidification power, and measurement of the rate of oxygen consumption. In practice, most brewers rely on the measurement of viability together with the quality-assurance approach of looking after stored yeast and ensuring that it is held in conditions under which it should maintain its quality.

Yeast Storage and Shipping

It is preferable that yeast be stored in a sanitized room that can be cleaned efficiently, is supplied with filtered air, and possesses a pressure higher than the surroundings (in order to impose an outward vector of contaminants). Ideally, it should be at or around 0°C. Even if storage is not in such a room, the tanks must be rigorously cleaned, chilled to 0–4°C, and have the facility for gentle rousing (mixing) to avoid hot spots. Yeast is stored in slurries ("barms") of 5–15% solids under 6 inches of beer, water, or KH_2PO_4. Some advocate the use of a little oxygenation at this stage to facilitate turnover metabolism of the yeast, although others think that oxygen should be rigorously excluded to avoid autolysis. An alternative procedure is to press the yeast and store it at 4°C in a cake form (20–30% dry solids). Pressed yeast may be held for about 10 days, slurries for 3–5 weeks.

Brewers seeking to ship yeast normally transport cultures for repropagation at the destination. However, greater consistency is achieved when it is feasible to propagate centrally and ship yeast for direct pitching. Such yeast must be contaminant-free, have high viability and vitality, be washed free from fermentable material, and be cold (0°C). The longer the distance, the greater the recommendation for low-moisture pressed cake.

Counting Yeast

Apart from the good condition of pitching yeast, it is also important that it is pitched in the correct quantity. The higher the pitching rate, the more rapid the fermentation. As the pitching rate increases, initially so too does the amount of new biomass synthesized, until at a certain rate, the amount of new yeast synthesized declines. The rate of attenuation and the amount of growth directly affect the metabolism of yeast and the levels of its metabolic products (i.e., beer flavor); hence the need for control.

Yeast can be quantified by weight or cell number. Typically, some 10^6 cells/ml are pitched for each degree Plato. At such a pitching rate, a lager yeast will divide four to five times in fermentation.

Yeast numbers can be measured using a hemocytometer, which is a counting chamber loaded onto a microscope slide. It is possible to weigh yeast or to centrifuge it down in pots that are calibrated to relate volume to mass, but in these cases it must be remembered that there are usually other solid materials present, e.g., trub.

Another procedure that has come into vogue is the use of capacitance probes that can be inserted in-line. An intact and living yeast cell acts as a capacitor and gives a signal, whereas dead ones (or insoluble materials) do not. The device is calibrated against a cell number (or weight technique) and therefore allows the direct read-out of the amount of viable yeast in a slurry. Other in-line devices quantify yeast on the basis of light scatter.

Further Reading

Stewart, G. G., and Russell, I. (1998) An Introduction to Brewing Science and Technology. Series III: Brewer's Yeast. The Institute of Brewing, London, UK.

12. Brewery Fermentations

Fermentation can be divided into two stages: primary and secondary (warm conditioning or maturation). The stages immediately before and after fermentation are wort collection and cold conditioning.

Wort Collection

Wort collection is the stage of the process during which the volume of cooled wort is measured. It may be performed in the primary fermentation vessel or in a separate tank. Traditional lager brewers use this stage to remove cold break, because they believe that this aids colloidal stability in the beer, circumvents the formation of sulfury flavors, and removes harsh bitter fractions derived from hops. The contrary argument is that the break acts as a nucleation site for CO_2 release during fermentation, which helps maintain a "rolling" fermentation and keep yeast in suspension and in contact with the wort it is fermenting.

In any brewery where fermentation of a single batch of wort is performed in several smaller vessels (e.g., Burton Union system), this is the stage at which wort and pitching homogeneity is ensured. However, there are real advantages in minimizing the numbers of tanks in breweries. Every extra vessel presents an additional cost in terms of
- capital
- wetting losses
- pumping losses
- cleaning costs
- maintenance costs
- space requirement
- volume of "work in progress"

For brewers convinced of the need to remove cold break, it is still possible to do this from the fermenter, e.g., by removing sediment from the base of cylindroconical fermenters 12–24 hr after filling or even by filtration or centrifugation.

Primary Fermentation

The primary fermentation is the stage proper in which yeast, through controlled growth, is allowed to ferment and wort to generate alcohol and the desired spectrum of flavors. The fermentation is regulated by control of several parameters:
- wort strength (°Plato)
- yeast pitching rate and viability

- oxygen
- temperature

Fermentation is monitored by measuring the decrease in specific gravity (alcohol has a much lower specific gravity than sugar; Fig. 82).

Ales are generally fermented at a higher temperature (15–20°C) than lagers (6–13°C), and therefore attenuation is achieved more rapidly. Thus, ale fermenting at 20°C may achieve attenuation gravity in 2 days, whereas lager fermented at 8.5°C may take 10 days. Remember that parameters such as temperature and higher yeast counts dictate the *rate* of fermentations, but it is the proportion of wort sugars that are fermentable, together with the presence of sufficient other nutrients (notably free amino nitrogen [FAN], oxygen, and zinc) that govern the *extent* of fermentation.

The temperature has a substantial effect on the metabolism of yeast, and thus the levels of a flavor substance like iso-butanol may be 16.5 and 7 mg/liter, respectively, for the ale and the lager.

Some brewers add zinc (e.g., 0.2 ppm) to promote yeast action and flocculation. Others use yeast foods (vitamin and amino acid supplements).

Fermentation is considered complete when the desired "attenuation" gravity has been reached.

During fermentation the pH falls, because yeast secretes organic acids and protons.

Surplus yeast will be removed at the end of fermentation, either by a process such as "skimming" from a traditional square fermenter employing top-fermenting yeast or from the base of a cone in a cylindroconical vessel. This is done to preserve the viability and vitality of yeast, but also to circumvent the autolysis and secretory tendencies of yeast that will be to the detriment of flavor and foam. There will still be sufficient yeast in the beer to effect the secondary fermentation.

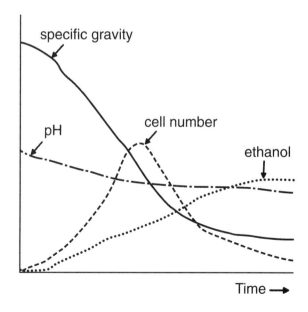

Fig. 82. Changes measurable during fermentation.

Secondary Fermentation

The "green" beer produced by primary fermentation needs to be "conditioned" to establish a desired carbon dioxide content and refinement of the flavor. Above all at this stage, an undesirable butterscotch flavor character caused by substances called vicinal diketones (VDKs; see later) needs to be removed.

Traditionally, lager beers fermented at lower temperatures have needed more prolonged maturation (storage: "lagering") in order to refine their flavor and develop carbonation. The latter depends on the presence of sugars, either those (perhaps 10%) that the brewer ensures are residual from the primary fermentation or those introduced in the "krausening" process. In the latter case, a proportion of freshly fermenting wort is added to the maturing beer. Krausen contains 8–9% (w/w) fermentable extract and a yeast count of approximately 40×10^6 cells/ml and is added at about 10% of the volume of the green beer.

Many brewers are unconvinced of the need for prolonged storage periods, and they tend to combine the primary and secondary fermentation stages. Once the target attenuation has been reached, the temperature is allowed to rise (perhaps by 4 degrees C), which permits the yeast to deal more rapidly with the VDKs. Carbonation will be achieved downstream by the direct introduction of gas.

Cold Conditioning

Once the secondary fermentation stage is complete, the temperature is dropped, ideally to $-1°C$ (or lower, short of freezing) to enable precipitation of materials that would otherwise cause a haze in the beer. The sedimentation of yeast is also promoted. This stage is most conveniently carried out in a vessel other than the fermenter.

Fermenters

Most fermenters these days are fabricated from stainless steel. This provides the necessary attributes of rigidity, ability to be shaped and welded, good thermal conductivity, chemical inertness (but beware of chloride), smooth surfaces to lessen the tendency for detritus to accumulate, and they are cost-effective. The gauges of metal needed for very large vessels are prohibitively expensive, so stainless steel is rolled onto mild steel as a support. The thermal conductivity of stainless steel is less than that of copper, but the greater rigidity of steel allows thinner gauges to be used, which facilitates heat transfer. Most vessels have heat-exchange panels in the walls, enabling brewers to dispense with less hygienic coils inside vessels.

Vessels will be lagged with materials that must be nontoxic, nonirritant, fire-resistant, inert, able to be sealed against moisture ingress, and, of course, effective thermal insulators. Most commonly, these materials are mineral wool, urethane insulants (but beware of toxic fumes in case of fire), and phenolic foams. They are usually covered with a sealed metal or plastic layer to prevent wetting and to make them look prettier.

The trend has been toward larger vessels, situated outside in order to minimize building costs with maximum output from minimum space.

Fundamentally, a fermentation vessel must feature
- clean and hygienic interior surfaces
- a cooling system
- a system for emptying at the base
- sampling systems
- a yeast-removal system
- the necessary capacity to hold the initial wort plus yeast, together with enough "freeboard" to accommodate the large amounts of foam that are produced during fermentation

Vessels may also need
- an in-built cleaning system
- an ability to be pressurized
- a CO_2-collection facility
- a rousing (mixing) facility

Modern vessels tend to be enclosed for microbiological reasons. However, across the world there remain a great many open tanks. Traditional systems include the Burton Union system (now restricted to a very few breweries) and the Yorkshire Square. Squares may typically hold 400 hl.

The bulk of the world's beer is fermented in cylindroconical vessels (Fig. 83). These were first described by L. Nathan in 1930. They can have a capacity of up to 13,000 hl. The following advantages have been listed for such vessels:
- They are relatively cheap to build, either inside or outside, in diverse aspect ratios.
- They are compact.
- The are flexible.
- Most brewing yeasts perform well in such vessels, although antifoam may be necessary, especially with top-fermenting strains. Antifoam is usually based on silicone and must be removed either with the yeast or on the filter if the foaming of the beer is not to suffer.
- Rousing can be achieved either mechanically, using inert gases, or by pumped circulation. The rousing system can be allied to further oxygen injection.
- There is a high speed of fermentation because of inherent turbulence. The CO_2 concentration is greater at the higher pressures at the base of the fermenter, leading to a gradient that forces a stream of liquid up the center of the fermenter and back down the walls.
- Wetting losses are low.
- CO_2 is easily collected.
- Yeast and cold break are easily removed. Most strains settle rapidly into the cone at the end of fermentation, particularly if cooled. Some breweries perform "cone-to-cone pitching," in which yeast is pumped directly from the cone of a fermenter where fermentation is complete to the cone of another fermenter containing fresh wort.
- There are minimum operator costs. Such vessels lend themselves to automation, manual operations amounting to little more than coupling up the vessel (either by opening valves with "hard" connections or screwing on the hose with "flexible" connections).

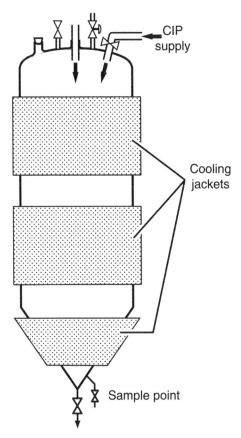

Fig. 83. Cylindroconical fermentation vessel. CIP = clean-in-place.

There are also disadvantages:
- There are longer fill times and decision processes regarding when to pitch the yeast.
- High CO_2 concentrations may develop, and for some beers these may subsequently need to be lowered
- Flavor-matching difficulties occur with some beers traditionally fermented in squares—hydrostatic pressures can be very different.
- Temperature stratification, especially at the end of primary fermentation, may amount to a difference of 2 degree C between the top of the fermenter and the main body. This higher temperature at the top may cause some shock excretion by yeast, with the generation of sulfury character.

Other types of fermenter include unitanks (similar to cylindroconical vessels but which include an agitator and increased cooling capacity to enable fermentation and cold conditioning in a single vessel), horizontal cylindrical tanks, and spherical tanks.

Cleaning-in-Place

Most plants in modern breweries are now cleaned using automatic clean-in-place (CIP) systems. In tanks, there are strategically placed spray balls that either rotate, ensuring an even firing of cleaning agents and rinsing water over the entire surface, or are fixed, giving a lower pressure stream directed at the roof of the vessel with the sides being cleaned by a cascade action. Pipe work is designed to avoid "dead legs" and the opportunity for cleaning streams to be forcefully driven through them with turbulent flow.

Cleaning cycles comprise
1. prerinse: removal of loose soil by water-rinsing
2. detergent wash: removal of soil by chemical action and heat
3. postrinse: removal of chemicals by clean water rinsing
4. sterilization: by chemicals or steam

Detergents are usually caustic-based (1% NaOH) and may contain sequestrants and chelators. The temperature may be 40–60°C. Detergents are recovered for reuse, and final rinse water may be recycled as prerinse water.

Acid detergents (phosphoric and/or nitric) may also be used. They are more expensive than caustic, but they produce a low pH effluent, which local regulations may require. They do not absorb CO_2 and can be used cold, both of which factors minimize the risk of vessels imploding, and they remove oxalate debris and hop resins more effectively than does caustic.

Sterilants may be hypochlorite-based or, increasingly, peracetic acid or chlorine dioxide.

High-Gravity Brewing

One opportunity to increase the output from a brewery is to maximize vessel utilization by fermenting wort at a higher gravity and then diluting the beer downstream with deaerated water to "sales gravity" (i.e., the required strength of the beer in package).

The technique depends, of course, on the facility to produce high-gravity worts. This is achieved by collecting only the strong worts in the first stages of wort separation and then diverting the weaker worts emerging on prolonged sparging to serve as the mashing-in water for the next brew (weak wort recycling). Additionally, gravity can be reinforced by the use of sugars in the kettle.

In the fermentation itself, it is necessary to increase the pitching rate pro rata to gravity, and higher oxygen levels will be used. Although there can be flavor problems at very high gravities, in practice such problems are not manifest at gravities up to 15°Plato.

Apart from capacity advantages, other benefits of high-gravity brewing include
- enhanced physical stability
- reduced energy costs and all other operational costs
- enhanced opportunity for control in blending to specification

Disadvantages are
- reduced brew house yield
- reduced hop utilization
- greater losses of concentrated materials
- greater stress on yeast
- poorer foams

Continuous Fermentation

Many industrial fermentations are performed continuously. With a solitary exception (Dominion Breweries in New Zealand), this is not the case for brewery fermentations, despite the obvious potential advantages for turnover and capacity. At times over the past 30 years, various breweries did install continuous fermentation processes, notably employing tower fermenters with upflow of the liquid stream through a heavily sedimentary yeast capable of forming a plug at the base of the vessel. By adjustment to the yeast content and the rate of wort flow, green beer could be produced in less than a day.

With that solitary exception, these fermenters have been stripped out: the main reasons given being inflexibility (most breweries produce a range of beers that demand diverse fermentation streams) and infection problems. It's bad enough having contamination in a batch fermenter, but it is substantially more inconvenient when the fermentation is continuous. There is also the matter of beer flavor: it is an undeniable truth that virtually any change in fermentation conditions, be they temperature, yeast concentration, or, in this case, continuous processing, leads to flavor shifts.

There is a resurgence of interest in continuous fermentation, including the use of so-called "immobilized yeast," where the yeast is attached to a solid support and the wort is flowed past. One Dutch brewer employs this type of process in the production of a low-alcohol beer, while others are experimenting with such fermentation systems for making full-strength beers on a boutique brewery scale. Furthermore, a brewer in Finland employs immobilized yeast in an accelerated process for eliminating VDKs at the end of fermentation (see later).

Further Reading

Boulton, C., and Quain, D. (2001) Brewing Yeast and Fermentation. Blackwell Science, London, UK.

13. Beer Flavor: Its Nature, Origins, and Control

The flavor of beer can be split into three separate components: *taste, smell* (*aroma*), and *texture* (*mouthfeel*).

Taste

There are only four proper tastes: sweet, sour, salt, and bitter. They are detected on the tongue. A related sense is the tingle associated with high levels of carbonation in a drink; this is due to the triggering of the trigeminal nerve by carbon dioxide. This nerve responds to mild irritants, such as carbonation and capsaicin (a substance largely responsible for the "pain delivery" of spices and peppers). Carbon dioxide also influences the extent to which volatile molecules are delivered via the foam and into the headspace above the beer in a glass.

The *sweetness* of a beer depends on its level of sugars, either those that have survived fermentation or those introduced as primings. The relative sweetness of a range of sugars is shown in Table 13.

Weight for weight, sucrose is two to three times sweeter than maltose. Sugars may also contribute to the body (mouthfeel) by increasing viscosity. So if sweetness as well as body is required, it is better to prime using a less sweet sugar that will need to be added in greater quantities, affording increased viscosity. Conversely, if body alone is required, a high dextrin content is needed, because dextrins (polymers containing in excess of four glucosyl units) have essentially no sweetness. However, it has been argued that unfeasibly large amounts of dextrin are needed for a significant impact on mouthfeel to be registered.

Sugars and dextrins covering a wide range of degrees of polymerization are now commercially available and are increasingly used by brewers keen to regulate the fermentability of their worts.

The amount of residual sugar surviving fermentation depends on the degree of fermentability of a wort (which, in turn, depends on the mashing temperature, with lower conversion temperatures promoting greater degrees of fermentability); the vigor and

Table 13. Relative sweetness of sugars

Sugar	Relative Sweetness
Glucose	0.7–0.8
Maltose	0.3–0.5
Fructose	1.1–1.2
Sucrose	1.0

Table 14. Relative bitterness of iso-α-acids

Compound	Typical Proportion in Beer (%)	Bitterness Rank[a]
trans-isocohumulone	7	1
cis-isocohumulone	30	2=
trans-isohumulone	10	2=
cis-isohumulone	40	4
trans-isoadhumulone	3	?
cis-isoadhumulone	10	?

[a] 4 indicates most bitter. Equal signs indicate identical bitterness.

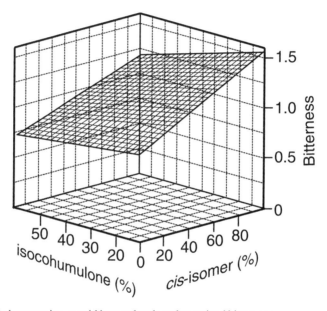

Fig. 84. Relationship between iso-α-acid isomer levels and perceived bitterness.

extent of fermentation (ergo yeast quality and quantity); and the degree to which a fermentation is allowed to proceed.

The principal contributors to *sourness* in beer are the organic acids produced by yeast during fermentation. These lower the pH: it is the H^+ ion imparted by acidic solutions that causes the sour character to be perceived on the palate. The acids concerned include acetic, lactic, and succinic. The more vigorous the fermentation, the more of these acids will be released by yeast.

Saltiness in beer is afforded by sodium and potassium. Of the anions present in beer, chloride and sulfate are believed by some to be of particular importance. Chloride contributes a mellowing fullness to a palate, while sulfate elevates the dryness of beer. Many brewers specify a chloride-sulfate ratio for their products. The levels of salts in beer are dependent on the methods used to purify brewing water, the level and type of salts used to bring this water into specification, and the contribution of ions from other raw materials that are added at various stages in the brewing process.

Table 15. Typical α-acid composition (%) of selected hop varieties

Variety	Cohumulone	Humulone	Adhumulone
Galena	36	51	13
Nugget	22	64	14
Wye Target	34	51	15

Table 16. Effect of mode of isomerization on yields (%) of *cis* and *trans* isomers of iso-alpha-acids

Means of Isomerization	*cis*	*trans*
Wort boiling	68	32
Aqueous alkali	55	45
MgO	80	20
Light	0	100

Bitterness is imparted by the iso-α-acids derived from the hop resins. There are six different iso-α-acids, and they differ in their relative bitterness (Table 14). These molecules differ not only in their bitterness but also in their ability to promote foaming.

Application of the time-intensity technique to the study of bitterness reveals that the chemical reduction of iso-α-acids (which increases their hydrophobicity and makes them less susceptible to breakdown to the compound responsible for the skunky flavor) renders these molecules substantially more bitter.

It is possible to construct three-dimensional plots that relate the balance of individual iso-α-acids to perceived bitterness (Fig. 84). Such diagrams can be used to predict the perceived bitterness from the balance of isomers present in any given beer. The proportions of iso-α-acid isomers found in a beer depend on the levels of the cohumulone, adhumulone, and humulone precursors in hops, which is variety dependent (Table 15) and on the mode by which they are isomerized (Table 16).

Smell

Many people believe that they can taste other notes in a beer. In fact, they are detecting them with the nose, the confusion arising because there is a continuum between the back of the throat and the nasal passages.

The smell (or aroma) of a beer is a complex distillation of the contribution of many individual molecules. No beer is so simple as to have its "nose" determined by one or even a very few substances. The perceived character is a balance between positive and negative flavor notes, each of which may be a consequence of one or a combination of many compounds of different chemical classes.

The *flavor threshold* is the lowest concentration of a substance that is detectable in beer. This is a simplistic factor, because compounds in a given class at concentrations well below their own flavor thresholds may cumulatively interact with a given aroma receptor site in the nasal system to afford a definite response by the taster. It is also apparent that, at a given concentration, a substance will be less detectable in a strong-impact, complex beer (e.g., a stout) than in a subtle, light beer (e.g., a low-carb lager).

A *flavor unit* is the concentration of a substance divided by its flavor threshold. Thus for dimethyl sulfide, which has a flavor threshold of approximately 30 parts per billion, two flavor units equates to 60 ppb. Compounds present in beer at one to two flavor units are weakly detectable, whereas at higher levels they should be readily perceived. The extent to which an increase in concentration of a substance leads to a more intense flavor perception differs for different substances (Fig. 85).

The substances that contribute to the aroma of beer are diverse. They are derived from malt and hops and by yeast activity (leaving aside for the moment the contribution of contaminating microbes). In turn, there are interactions between these sources, such as yeast converting one flavor constituent from malt or hops into a different one.

Alcohols

Various alcohols influence the flavor of beer (Table 17). The most important of these is ethanol, which is present in most beers at levels at least 350-fold higher than any other

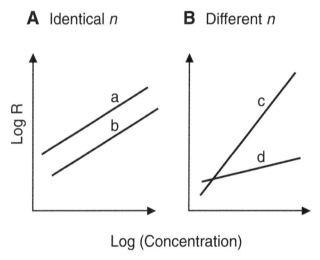

Fig. 85. Stevens' power function. Stevens' power law describes the relationship between a sensory response and concentration of an aroma-active compound: $R = kC^n$, where R is the sensory response, C is concentration, k is a constant, and n is the psychophysical constant. The slope of the line is given by n. At any given concentration of compound a, it generates a more intense response than compound b. However, the extent to which an increase in concentration registers as an impact on the subject's response is identical. The slope of compound c is much greater than that of compound d; i.e., smaller increments in level of c lead to a perceptible change in sensory response than would be the case for d.

Table 17. Some alcohols in beer

Alcohol	Flavor Threshold (mg/liter)	Perceived Character
Ethanol	14,000	Alcoholic
Propanol	600	Alcoholic
iso-Butanol	100	Alcoholic
iso-Amylalcohol	50	Alcohol, banana, vinous
Tyrosol	200	Bitter
Phenylethanol	40–100	Roses, perfume

alcohol. Ethanol contributes directly to the flavor of beer, registering a warming character, although the textbooks refer to the flavor imparted by alcohols as alcoholic!

Ethanol also influences the flavor contribution of other volatile substances in beer. Because it is usually quantitatively second only to water as the main component of beer, it is not surprising that it moderates the flavor impact of other substances. It does this by affecting the vapor pressure of other molecules (i.e., their relative tendency to remain in beer or to migrate to the headspace of the beer).

The higher alcohols in beer are important as the immediate precursors of the esters, which are proportionately more flavor active. Therefore, it is important to be able to regulate the levels of the higher alcohols produced by yeast if ester levels are also to be controlled.

The pathway for the production of ethanol was described earlier. The higher alcohols are produced during fermentation by two routes: "catabolic" and "anabolic."

In the catabolic (Ehrlich) route, yeast amino acids taken up from the wort by yeast are transaminated to α-keto acids, which are decarboxylated and reduced to alcohols:

$$RCH(NH_2)COOH + R^1COCOOH \rightarrow RCOCOOH + R^1CH(NH_2)COOH$$
$$RCOCOOH \rightarrow RCHO + CO_2$$
$$RCHO + NADH + H^+ \rightarrow RCH_2OH + NAD^+$$

The anabolic route starts with pyruvate (the end-point of the EMP pathway proper), the higher alcohols being "side shoots" from the synthesis of the amino acids valine and leucine (Fig. 86). The penultimate stage in the production of all amino acids is the formation of the relevant keto acid, which is transaminated to the amino acid. Should there be conditions where the keto acids accumulate, they are then decarboxylated and reduced to the equivalent alcohol (cf., Ehrlich pathway). Essentially, therefore, the only

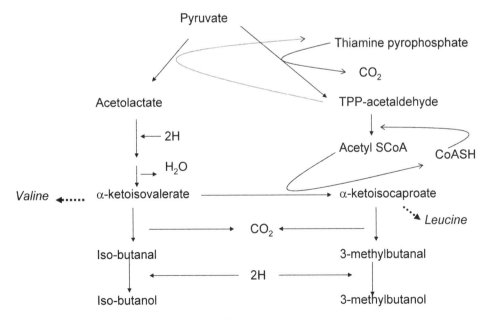

Fig. 86. Biosynthetic pathway for the production of higher alcohols in yeast.

difference between the pathways is the origin of the keto acid: either the transamination product of an amino acid assimilated by the yeast from its growth medium or synthesized de novo from pyruvate.

In view of the above, it is unsurprising that the levels of free amino nitrogen (FAN) in wort influence the levels of higher alcohols formed. Higher alcohol production is increased at both excessively high and insufficiently low levels of assimilable nitrogen available to the yeast from wort. If levels of assimilable N are low, then yeast growth is limited and there is a high incidence of the anabolic pathway. If levels of N are high, then the amino acids feed back to inhibit further synthesis of them and the anabolic pathway becomes less important. However, there is a greater tendency for the catabolic pathway to "kick in." The level of FAN obviously depends on the grist composition (extent of modification of malt, level of N-free adjunct, etc.). A FAN level of 140–150 mg/liter for wort of 10°Plato has been recommended. The impact of the C:N ratio on the levels of various flavor compounds in beer (see later) is tabulated in Table 18.

Even more important than FAN levels, though, is the yeast strain: more of these compounds are produced by ale strains than by lager strains.

Fermentations at higher temperatures increase the production of higher alcohols. Conditions favoring increased yeast growth (e.g., excessive aeration or oxygenation) promote higher alcohol formation, but this can be countered by application of a top pressure on the fermenter. The reasons that increased pressure has this effect are unclear, but it has been suggested that it may for some reason be due to an accumulation of carbon dioxide. Beer produced in different sizes and shapes of vessel, displaying different hydrostatic pressures, vary in their content of higher alcohols (and therefore of esters). This can be a problem for product matching between breweries (e.g., in franchise brewing operations).

Esters

Various esters may make a contribution to the flavor of beer (Table 19). Of these, most brewers worry about ethyl acetate and iso-amyl acetate, probably with some justification.

Table 18. Impact of sugar:amino acid ratio in wort on flavor volatile production by yeast

Flavor Compound	Flavor Impact	Impact of C:N Ratio
Dimethyl sulfide	Sweet corn	Higher ratio → more DMS
Esters, e.g., iso-amyl acetate	Banana	Higher ratio → more ester
Higher alcohols, e.g., the methylbutanols	Solvent	Too low or too high a ratio → more alcohol
Vicinal diketones, e.g., diacetyl	Butterscotch	Higher ratio → more VDK
Organic acids, e.g., citric	Sour	Higher ratio → lower pH through reduced buffering
Fatty acids, e.g., decanoic	Various	Higher ratio → less fatty acid

Table 19. Some esters in beer

Ester	Flavor Threshold (mg/liter)	Perceived Character
Ethyl acetate	33	Solvency, fruity, sweet
Iso-amyl acetate	1.0	Banana
Ethyl caprylate	0.9	Apples, sweet, fruity
Phenylethyl acetate	3.8	Roses, honey, apples

The esters are produced from their equivalent alcohols (ROH) through catalysis by the enzyme alcohol acetyl transferase (AAT), with acetyl-coenzyme A ($CH_3COSCoA$, which is the key form of "active acetate" in the yeast cell and which is formed from pyruvate in the pyruvate dehydrogenase reaction) being the donor of the acetate group:

$$ROH + CH_3COSCoA \rightarrow CH_3COOR + CoASH$$

Clearly the amount of ester produced depends on the levels of acetyl CoA, of alcohol, and of AAT. Esters are formed under conditions when the acetyl CoA is not required as the prime building block for the synthesis of key cell components. In particular, acetyl CoA is the start point for the synthesis of lipids, which the cell requires for its membranes. Thus, factors promoting yeast production (e.g., high levels of aeration/oxygenation) *lower* ester production and vice versa.

Yeast strain is probably more important, because some strains are more predisposed to generating esters than others. This may relate to the amount of AAT they contain. The factors that dictate the level of this enzyme present in a given yeast strain are not fully elucidated, but it does seem to be present in raised quantities when the yeast encounters high-gravity wort, and this may explain the disproportionate extent of ester synthesis under these conditions. There is no substantial problem up to a gravity of 15°Plato, with the ester levels in the diluted beer not being excessively raised. It is at higher gravities that the problems arise. This explains also the propensity for beers such as barley wines to have a very estery character. Going from 10 to 20°Plato increases the ester level at least fourfold. It can be countered by increasing the amount of oxygen available to the yeast. Ester formation may also be reduced by increasing the level of lipid in the wort, e.g., by use of "dirty worts."

Vicinal Diketones

Whereas the esters and higher alcohols often make positive contributions to the flavor of beer, few beers (with the possible exception of some ales) warrant the presence of the vicinal diketones (VDKs) diacetyl and (less importantly) pentanedione (Table 20).

Elimination of VDKs from beer depends on the fermentation process being well run. These substances are offshoots of the pathways by which yeast produces the amino acids valine and isoleucine (and therefore there is a relationship to the anabolic pathway of higher alcohol production).

The pathway for diacetyl production is shown in Figure 87 because it is more significant (in respect to diacetyl being present at higher levels and with a lower flavor threshold).

The precursor molecules leak out of the yeast and break down spontaneously to form VDKs. Happily, the yeast can mop up the VDKs, *provided* it remains in contact with the beer and is in good condition:

Table 20. Vicinal diketones in beer

Vicinal Diketone	Flavor Threshold (mg/liter)	Perceived Character
Diacetyl	0.1	Butterscotch
Pentanedione	0.9	Honey

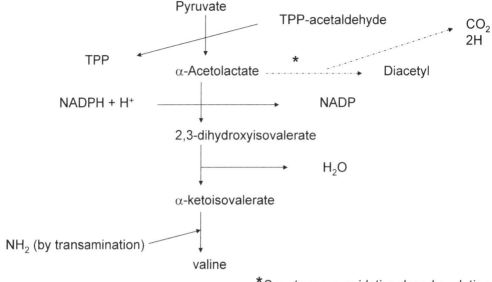

Fig. 87. Origins of diacetyl in brewing yeast.

$$CH_3COCOCH_3 + NADH + H^+ \rightarrow CH_3CH(OH)COCH_3 + NAD^+$$
$$CH_3CH(OH)COCH_3 + NADH + H^+ \rightarrow CH_3CH(OH)CH(OH)CH_3 + NAD^+$$

Reductases in the yeast reduce diacetyl successively to acetoin and 2,3-butanediol, both of which have much higher flavor thresholds than diacetyl. Analogous reactions occur with pentanedione.

Many brewers allow a temperature rise at the end of fermentation to facilitate more rapid removal of VDKs. Others introduce a small amount of freshly fermenting wort later on as an inoculum of healthy yeast (krausening, see earlier). In Finland, the Synebrychoff Company accelerates the decomposition of the precursors by heating beer (in the strict absence of oxygen, otherwise cooking will occur) prior to use of an immobilized yeast to consume the free VDKs. Alternatively, an enzyme (acetolactate decarboxylase) can be added to the fermenter, with the result that acetolactate is converted directly to acetoin without the intermediacy of the much more flavor potent diacetyl.

Persistently high diacetyl levels in a brewery's production may be indicative of an infection by *Pediococcus* or *Lactobacillus* bacteria. A disproportionately high ratio of diacetyl to pentanedione indicates that there is an infection problem.

Figure 88 illustrates the common origins of alcohols, esters, VDKs, acetaldehyde, and fatty acids.

Sulfur Compounds

In many ways, the most complex flavor characters in beer are caused by the sulfur-containing compounds. There are many of these in beer (Table 21), and they make various contributions.

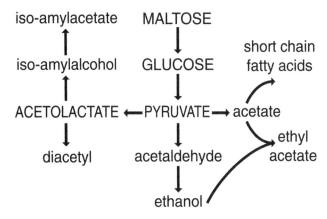

Fig. 88. Interrelationships among routes of flavor-compound production by yeast.

Table 21. Some sulfur-containing substances in beer

S-Containing Compound	Flavor Threshold (mg/liter)	Perceived Character
Hydrogen sulfide	0.005	Rotten eggs
Sulfur dioxide	25	Burnt matches
Methyl mercaptan	0.00015	Rotting vegetables
Ethyl mercaptan	0.0005	Rotting vegetables
Propyl mercaptan	0.0005	Onion
Dimethyl sulfide	0.03	Sweet corn
Dimethyl disulfide	0.0075	Rotting vegetables
Dimethyl trisulfide	0.00001	Rotting vegetables, onion
Methyl thioacetate	0.05	Cooked cabbage
Diethyl sulfide	0.0012	Cooked vegetables, garlic
Methional	0.04	Cooked potato
3-Methyl-2-butene-1-thiol	0.000004–0.001	Light-struck, skunky

Some ales have a deliberate hydrogen sulfide character, not too much, but just enough to give a nice "eggy" nose.

Lagers tend to have a more complex sulfury character. Some lagers are relatively devoid of any sulfury nose. Others, though, have a distinct dimethyl sulfide (DMS) character, while some have characters ranging from cabbagy to burnt rubber. This range of characteristics renders substantial complexity to the control of sulfury flavors.

All of the DMS in a lager ultimately originates from a precursor, S-methylmethionine (SMM), which is produced during the germination of barley (Fig. 89). SMM is heat-sensitive and is broken down rapidly whenever the temperature gets above about 80°C during the process. Thus, SMM levels are lower in the more intensely kilned ale malts and, as a result, DMS is a character more associated with lagers. SMM leaches into wort during mashing and is further degraded during boiling and in the whirlpool. If the boil is vigorous, most of the SMM is converted to DMS and driven off. (The half life of SMM at 100°C is 38 min, and this value doubles for every 6 degree C decrease in temperature.) In the whirlpool, conditions are gentler than in the kettle: any SMM surviving the boil will be broken down to DMS, but the DMS tends to stay in the wort. Brewers seeking to retain some DMS in their beer specify a finite level of SMM in their malt and manipulate

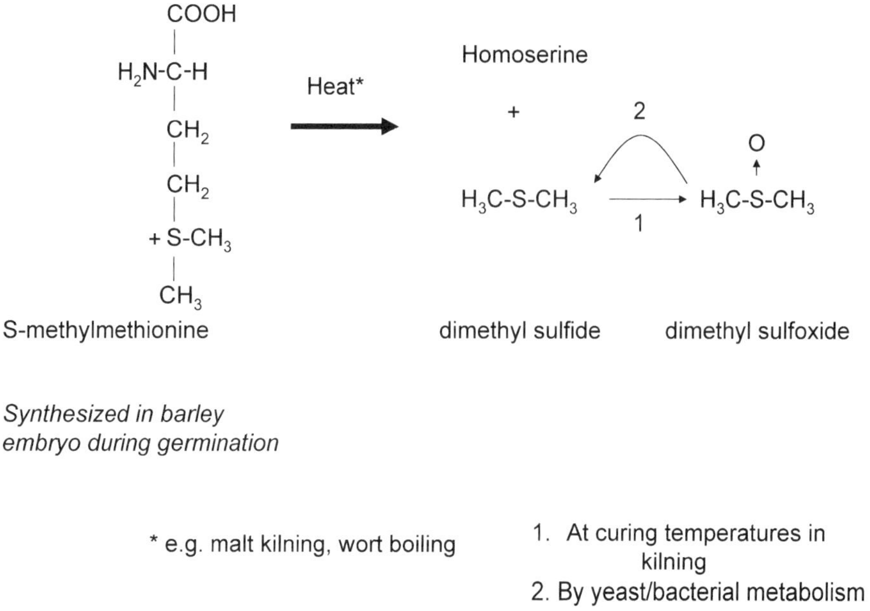

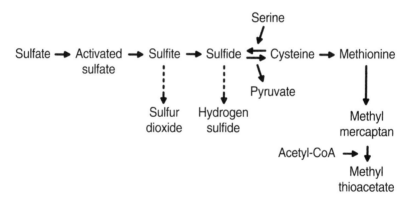

Fig. 89. Sources of dimethyl sulfide in beer.

Fig. 90. Sources of some other S-containing compounds in beer.

the boil and whirlpool stages in order to deliver a certain level of DMS into the pitching wort. During fermentation, much DMS will be lost with the gases, so the level of DMS required in the wort will be somewhat higher than that specified for the beer. Another complication is that some of the SMM is converted during kilning into a third substance, dimethyl sulfoxide (DMSO), which is not heat labile but is water soluble. It gets into wort at high levels, and some yeasts are quite adept at converting it to DMS.

If phenylethanol and phenylethyl acetate are present in beer, they tend to suppress the perception of DMS. Why this is the case is so far a mystery, but it is almost certainly the case that there are many such antagonisms that influence the perception of many of the flavor characters in beer.

Hydrogen sulfide (H_2S) can be produced by more than one pathway in yeast (Fig. 90). It may be formed by the breakdown of amino acids such as cysteine or peptides like glutathione or by the reduction of inorganic sources such as sulfate and sulfite. Yeast strain has a major effect on the levels of H_2S that are produced during fermentation. For all strains, more H_2S will be present in green beer if the yeast is in poor condition, because a vigorous fermentation is needed to purge H_2S. Any other factor that hinders fermentation (e.g., a lack of zinc or vitamins) will also lead to an exaggeration of H_2S levels in beer. Furthermore, H_2S is a product of yeast autolysis, which is more prevalent in unhealthy yeast. It is also worth noting that healthy yeast in products such as cask-conditioned ales might also produce H_2S by reducing the sulfite preservative in finings.

How to control the other sulfury characters is even less clear. The routes by which most of them are formed are unknown, making this one of the most fertile grounds for future research. Recently it was shown that the enzyme that produces methyl thioacetate is the same one that makes the oxygen-containing esters (i.e., AAT).

When iso-α-acids are exposed to light, they break down, react with sulfur sources in the beer, and form a substance called 3-methyl-2-butene-1-thiol (MBT), which has an intense skunky character and is detectable at extremely low concentrations. There are two ways of protecting against this: do not expose beer to light or increase bitterness by using chemically modified bitter extracts, the reduced iso-α-acids. These acids are very effective, but it is critical that they are the sole source of bitterness for light-resistant beers. For instance, yeast from a brew in which conventional hopping was employed to ferment a beer will bind trace levels of unreduced bitter compounds that are sufficient to lead to a skunky flavor.

Hop Aroma

(Also see Chapter 9.) The point at which hops are added during beer production determines the resulting flavor that they impart. The practice of adding aroma hops close to the end of boiling (late hopping) still results in the substantial evaporation of volatile material, but of the little that remains, much is transformed into other species (e.g., the hop oil component humulene can be converted to the more flavor-active humulene epoxide). Further changes then occur during fermentation, such as the transesterification of methyl esters to their ethyl counterparts. The resultant late hop flavor is rather floral in character and is generally an attribute associated more with lager beers.

In a generally distinct practice, hops may be added to the beer right at the end of production. This process of dry hopping gives certain ales their characteristic aroma. The hop oil components contributed to beer by this process are very different from those from late hopping, with mono- and sesquiterpenes surviving generally unchanged in the beer. The result is an aroma more similar to that of raw hops, generating a resinous note in the final product.

Sulfur-containing compounds have also been found in hop oil fractions. In particular, these tend to be sulfur derivatives of terpenes, such as caryophyllene episulfide, and disulfide adducts of myrcene and probably result from the burning of native sulfur in hop kilns (as has been practiced to counter certain infections). Hops have even been reported to contribute as much as 15 µg/liter of DMS to finished beer.

Malt-Derived Flavors

Apart from the contribution of much of the DMS character to beer, malt can make many other contributions, most of which are as yet poorly defined chemically. Only

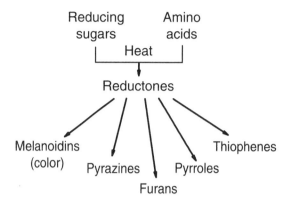

Fig. 91. Maillard reaction products.

relatively recently have studies led to the establishment of robust tasting procedures for malts that have enabled a start to be made in understanding the influence of malting conditions on flavor delivery from malts. In terms of interpreting these observations chemically, there is much still to do.

Malty character is due in part at least to isovaleraldehyde, which is formed by a reaction between one of the amino acids (leucine) and reductones in the malt. The toffee and caramel flavors in crystal malts and the roasted, coffee-like notes found in darker malts are caused by various complex components generated from amino acids and sugars that cross-react during kilning. Another key change during kilning is the disappearance of grassy notes.

The impact of the Maillard reaction, which is responsible for color and flavor formation in brewing and beer, is shown in Figure 91.

Other Flavors

Acetaldehyde, which is the immediate precursor of ethanol in yeast, has a flavor threshold of between 5 and 50 mg/liter and imparts a "green apples" flavor to beer. High levels should not survive into beer in successful fermentations, because yeast will efficiently convert the acetaldehyde into ethanol. Persistently high levels are an indication of premature yeast separation, poor yeast quality, or a *Zymomonas* infection.

The *short-chain fatty acids* (Table 22) are made by yeast as intermediates in the synthesis of the lipid membrane components. For this reason, the control of these acids is exactly analogous to that of the esters (see earlier). If yeast needs to make less lipid (under conditions where it needs to grow less), then short-chain fatty acids will accumulate.

Some beers (e.g., some wheat beers) feature a *phenolic* or clove-like character. This is due to molecules of substances such as 4-vinylguaiacol (4-VG), which is produced by certain *Saccharomyces* species, including *S. diastaticus*. Its unwanted presence in a beer is an indication of a wild yeast infection. 4-VG is produced by the decarboxylation of ferulic acid (a component of barley endosperm cell walls) by an enzyme present in the strains used to make wheat beers and *S. diastaticus* and other wild yeasts, but it is not present in brewing strains.

Table 22. Some short-chain fatty acids in beer

Fatty Acid	Flavor Threshold (mg/liter)	Perceived Character
Acetic	2175	Vinegar
Propionic	150	Acidic, milky
Butyric	2.2	Cheesy
3-Methylbutyric	1.5	Sweaty
Caproic	8	Vegetable oil
Caprylic	15	Goaty
Phenylacetic	2.5	Honey

Another undesirable character is *metallic*, which, if present in beer, is most likely to be due to the presence of high levels (greater than 0.3 ppm) of iron. One known cause is the leaching of the metal from the filter aid: if water is used to slurry the filter aid for precoating, the first beer subsequently put through the filter will extract a high level of iron.

Mouthfeel

Any drinker who has ordered a beer containing nitrogen gas will appreciate that one can talk of the mouthfeel and texture of beer. Nitrogen not only imparts a tight, white head to a beer, but it also gives rise to a creamy texture. However, the partial replacement of carbon dioxide with nitrogen gas suppresses several beer flavor attributes, such as astringency, bitterness, and hop aroma, and reduces the carbon dioxide "tingle." Other components of beer, such as the astringent polyphenols, may also play a part. Physical properties, such as viscosity, are influenced by residual carbohydrate in the beer and might also contribute to the overall mouthfeel of a product. It is thought that turbulent flow of liquids between the tongue and the roof of the mouth results in increased *perceived* viscosity and therefore enhanced mouthfeel.

Indirect Effects on Beer Flavor

A customer's perception of flavor is incontrovertibly related to the appearance of a beer. The presence or absence of a head on the beer, the color, the clarity, and even the presence of scuffing or a torn label on a bottle can all affect the perceived flavor of a beer. Achieving the desired flavor quality in a product demands attention to more than the flavor components alone.

Further Reading

MacDonald, J., Reeve, P. T. V., Ruddlesden, J. D., and White, F. H. (1984) Pages 47–198 in: Progress in Industrial Microbiology, vol. 19. M. E. Bushell, ed. Elsevier, London, UK.

14. Downstream Processing: Cold Conditioning, Filtration, and Stabilization

When a beer leaves the fermenter, it is not the finished article. It is highly unlikely to be sufficiently clear or "bright" and will certainly contain substances that will come out of solution in the package. Its flavor may still require some refining. All brewers recognize the need to attend to the "raw" or "green" beer, but they differ in their opinions about how intense and involved this processing needs to be.

Flavor Changes During the Aging of Beer

As we saw earlier, a rate-limiting step for moving beer onward from the fermenter is the time taken to mop up diacetyl and its precursor. Some people refer to this as "warm conditioning." Many brewers consider this to mark the end of the useful flavor changes that they can dictate in the brewery. The traditionalists contend that the beer still needs to be stored. There is, however, very little published data to indicate what, if any, further changes take place in the flavor of beer when it is aged in the brewery.

Some major brewing companies insist on holding lager for a prolonged period at low temperatures (decreasing from 5 to 0°C). This process (called lagering) is left over from prerefrigeration days, when the removal of bottom-fermenting yeast demanded that the beer be held for a long time, with chilling perhaps facilitated by blocks of ice. Traditionally, beer from an already relatively cool fermentation (<10°C) was run to a cellar at a stage when there was still about 1% fermentable sugar and sufficient yeast left in it. The yeast would consume traces of potentially destabilizing oxygen and, by fermenting the sugar, release carbon dioxide that would remain in solution to a greater extent at the lower temperatures. In this way, the beer might be held at 0°C for perhaps 50 days or even longer. Yeast would settle out by the end of this time, together with protein and other material that otherwise would "drop out" as an unsightly haze in the finished beer in the customer's glass.

These days, the technology exists to cover all these requirements for prolonged storage, including the use of clarifying agents, filters, stabilizing agents, and carbonation systems, all allied to the use of refrigeration.

Clarification of Beer

Cold Conditioning

Two types of particles need to be removed from beer at the end of fermentation: yeast and cold break. In addition, substances present in solution at this stage, but that tend to form particles when beer is in the trade, must also be eliminated.

The first mechanism by which particles separate from beer is simple gravitational pull. Most brewers ensure that their beer is chilled to either −1 or −2°C after it has achieved the degree of fermentation and maturation that they deem it requires. Particles progressively sediment at this temperature in proportion to their size, and materials will be brought out of solution, substances that might otherwise emerge as unsightly haze in the packaged beer.

The equation (Stokes' Law) governing sedimentation is

$$v_g = \frac{x^2 (d_p - d_l) g}{18\eta}$$

where v_g is the terminal settling velocity of a particle under gravitational force
    x is the diameter of the particle
    d_p is the density of the particle
    d_l is the density of the liquid
    g is the acceleration due to gravity
    η is the viscosity of the liquid

Obviously the rate of particle sedimentation can be increased by maximizing the density difference between particles and the liquid, increasing the gravitational force, reducing the viscosity of the liquid, and, in particular (because it is a squared term) increasing particle size.

The acceleration due to gravity can be increased by using a centrifuge, and many brewers use a centrifuge en route to the cold tank, taking great precautions not to cause shear damage or allow excess oxygen pick-up.

To facilitate the sedimentation of particles, some brewers add isinglass finings. These are solutions of collagen derived from the swim bladders of certain species of fish from the South China Sea region. Collagen has a net positive charge at the pH of beer, whereas yeast and other particulates have a net negative charge. Opposite charges attracting, the isinglass forms a complex with these particles, and the resultant large agglomerates sediment readily because of an increase in particle size. Sometimes the isinglass finings are used alongside "auxiliary finings" based on silicate, the combination being more effective than isinglass alone.

One brewer uses wood chips. Over the years, these chips have been mostly derived from well-seasoned beech and, individually, are a few inches wide and as much as a foot long. They therefore present a very ample surface area upon which insoluble materials, including yeast cells in maturing beer, can stick.

Filtration

After a minimum period of typically three days in this "cold conditioning," the beer is generally filtered. Diverse types of filters are available. Although crossflow and membrane filtration are finally assuming economic viability, most beer filters still feature powder filtration through various designs of filter: plate and frame, vertical leaf, horizontal leaf, etc. We need not discuss their respective designs or pros and cons. Suffice to say that they provide a surface against which the filter aid is able to accumulate as it removes particles from beer.

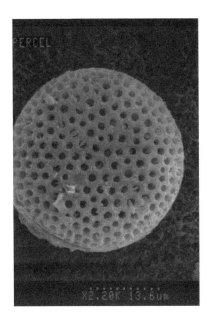

Fig. 92. Kieselguhr.

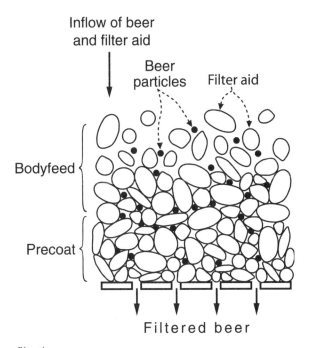

Fig. 93. Principle of beer filtration.

The filter aid both traps particles and prevents the system from clogging. Two major kinds of filter aid are in regular use: kieselguhr (diatomaceous earth) and perlite. The former comprises fossils or skeletons of primitive organisms called diatoms (Fig. 92). These can be mined and classified to provide grades that differ in their permeability. Particles of kieselguhr contain pores into which other particles (such as those found in beer) can pass, depending on their size. Unfortunately, there are health concerns associated with kieselguhr; inhalation of its dust adversely affects the respiratory tract.

Perlites are derived from volcanic glasses crushed to form microscopic flat particles. They are better to handle than kieselguhr but are not as efficient as filter aids.

Filtration starts when a precoat of filter aid is applied to the filter by cycling a slurry of filter aid through the plates (Fig. 93). This precoat is generally of quite a coarse grade, whereas the filter aid (the body feed), which is dosed into the beer during the filtration proper, tends to be a finer grade. It is selected according to the particles within the beer that need to be removed. If a beer contains a lot of yeast but relatively few small particles, then a relatively coarse grade is best. If the converse applies, then a fine grade will be used.

The principles of beer filtration are similar to those we encountered when considering lautering (Darcy's Law). Long filtration runs depend on the conservative application of pressure and are easier to achieve if factors such as viscosity are low. As viscosity is substantially increased at lower temperatures and as beer should be filtered at as near to cold stabilization temperature as possible, it is particularly beneficial if substances like β-glucan are removed prior to this stage.

Stabilization of Beer

In addition to filtration, various other treatments may be applied to beer downstream, all with the aim of enhancing the shelf life of the product. There are three principal ways in which beer can deteriorate with time: by staling, by throwing a haze, and by becoming infected.

The flavor of beer changes in various ways in the package. The most significant of these changes are due to oxidation. It is now believed that oxidation reactions can take place throughout the brewing process, and that the tendency to stale may be built into a beer long before it is packaged and dispatched to trade. However, no brewer would argue against the fact that the oxygen level in the beer in its final container should be as low as possible.

The freshly filtered beer, called "bright beer," should have an oxygen content below 0.1 ppm. In part, this is achieved by running the beer from the filter into a tank that has been equilibrated in carbon dioxide or even nitrogen. The flow of beer into the vessel will be gentle. And if, once the vessel is full, the oxygen content of the beer exceeds specification, the vessel will be purged with carbon dioxide or nitrogen to drive off the surplus oxygen. Some brewers add antioxidants at this stage, such as sulfur dioxide or ascorbic acid (vitamin C), but the latter is of dubious merit.

A haze in beer can be due to various materials, but principally it is due to the cross-linking of certain proteins and certain tannoids (polyphenols) in the product. Therefore, if one or both of these materials is removed, the shelf life is extended.

As we have already seen, the brew house operations are in part designed to precipitate out protein-polyphenol complexes. Thus, if these operations are performed efficiently, then much of the job of stabilization is achieved. Good, vigorous, rolling boils, for instance, ensure precipitation. Before that, avoidance of the last runnings in the lautering operation prevent excessive levels of tannoid entering the wort.

We have also seen that cold conditioning has a major role to play by chilling out protein-polyphenol complexes, enabling them to be taken out on the filter. Control over oxygen and oxidation is just as important for colloidal stability as for flavor stability, because it is particularly the oxidized polyphenols that tend to cross-link with proteins.

For really long shelf lives, though, and certainly if the beer is being shipped to extremes of climate, additional stabilization treatments are necessary, as described below.

Gas Control

In addition to stabilization downstream, final adjustment is made to the level of gases in the beer. As we have seen, it is important that the oxygen level in the bright beer be as low as possible. Unfortunately, whenever beer is moved around and processed in a brewery, there is the risk of oxygen pick-up. For example, oxygen can enter through leaky pumps. A check on oxygen content is made once the bright beer tank is filled, and if the level is above specification (which most brewers set at 0.1 ppm or less), oxygen has to be removed. This is achieved by purging the tank with an inert gas, usually nitrogen, from a sinter in the base of the vessel. It is not a desirable practice, because whenever a purging process takes place there is a foaming on beer. The foam sticks to the side of the tank and dries, the resulting flakes falling into the beer to form unsightly bits.

The level of carbon dioxide in a beer may need to be either increased or decreased. The majority of beers contain between 2 and 3 volumes of CO_2, whereas most brewery fermentations generate and retain "naturally" no more than 1.2–1.7 volumes of the gas. The simplest and most usual procedure by which CO_2 is introduced is by injection as a flow of bubbles as beer is transferred from the filter to the bright beer tank.

If the CO_2 content needs to be dropped, this is a more formidable challenge. It may be necessary for beers that are supposed to have a relatively low carbonation (beers such as the "nitrokegs" or draft-beers-in-can), and oxygen reduction can be achieved by purging. However, concerns about "bit" production have stimulated the development of gentle, membrane-based systems for gas control. Beer is flowed past membranes made from polypropylene or polytetrafluoroethylene, membranes that are water-hating and therefore don't "wet-out." Gases, but not liquids, pass freely across such membranes, the rate of flux being proportional to the concentration of each individual gas and dependent also on the rate at which the beer flows past the membrane. If the CO_2 content on the other side of the membrane is lower than that in the beer, the level of carbonation in the beer will decrease. If the CO_2 content on the other side of the membrane is higher than that in the beer, the beer will become more highly carbonated. Gases behave independently, so the membranes can be used simultaneously to remove CO_2 and oxygen from a beer, providing the levels of both gases are lower on the other side of the membrane. This technique is also an excellent opportunity to introduce nitrogen into beer.

Recovered Beer

There are several points at which beer is "held up" in the process, prime among which are
- with yeast recovered from a fermenter
- with solids recovered in cold conditioning

The brewer may choose to recover this beer, which would otherwise be wasted and would also contribute to effluent costs. Yeast may be washed, pressed, and squeezed on yeast presses (rather like the mash filters).

It is essential that any recovered beer is of high quality (e.g., no flavor taints, low O_2). It is likely to be flash pasteurized before return to the process stream (to avoid a microbiological risk). It may be added back at any convenient stage from fermenter onward, providing that quality is not jeopardized.

Further Reading

Munroe, J. H. (1995) Aging and finishing. Pages 355-379 in: Handbook of Brewing. W. A. Hardwick, ed. Dekker, New York.

15. Haze Instability

Beer is prone to instability:
- haze instability, also called colloidal or nonbiological instability—the tendency for beers to become turbid or develop bits or sediments
- flavor instability—the tendency of beer to change its flavor in the package
- biological instability—the tendency of beer to become infected with bacteria and yeasts
- light instability—the tendency of beer to develop unacceptable flavor when exposed to light
- foam instability—the tendency of beer to have a less-good foam than desired by producer and consumer
- gushing—when beer surges uncontrollably out of a can or bottle when opened

In this section, we will focus on haze and other forms of turbidity in beer.

Types of Haze

Bits can be formed in beers subjected to agitation, e.g., when shipped long distances. These bits, which contain protein and perhaps pentosans, are thought to arise as the skins around foam generated within the package.

Bits may also arise through the interaction of different stabilizing agents in beer, e.g., cross linking of papain with propylene glycol alginate (PGA), which can occur during pasteurization. PGA can also bind to residual isinglass finings to throw gelatinous precipitates when beer encounters high temperatures.

"Invisible" hazes, which are more accurately called "pseudo" hazes, are very small particles (<0.1 μm) that scatter light strongly at 90° to incident. Such particles may comprise unmodified regions of the starchy endosperm of barley (retrograded α-glucan) or disintegrating yeast.

Visible hazes in beer have been reported as resulting from
- cross-linking of protein (from the grist) with polyphenols (from malt or hops)
- starch that has not been properly broken down in the brew house
- pentosans from wheat-based adjunct
- oxalate from calcium-deficient worts
- β-glucan from inadequately modified malt
- carbohydrate and protein from damaged yeast
- lubricants from can lids
- dead bacteria from malt

The composition of haze can be established by staining procedures and by chemical and enzymic tests. The latter are more reliable, as the underlying cause of a haze may be at the nucleus of the particle rather than the material that resides on the surface.

Protein-Polyphenol Hazes

When a low molecular weight polyphenol cross-links with a protein through weak interactions such as hydrogen bonds, a chill haze is produced. (Chill haze—a haze that forms when beer is chilled to 0°C but that returns to solution when the beer returns to 20°C.) Particle sizes are 0.1–1.0 µm.

Polyphenols, however, are prone to polymerization (by oxidation), and when they interact with protein after this process they form a permanent haze, with particles of 1–10 µm. (Permanent haze—a haze that is present in beer even at 20°C.) If chill haze is dealt with by the brewer (cf., the cold conditioning stage), then the beer will not go on to develop a permanent haze.

Haze-Active Proteins

As few as 2 mg of protein per liter is sufficient to induce a haze of 1 European Brewing Convention (EBC) unit. As most beers contain in excess of 2 g of protein per liter, it is apparent that there is vastly more protein in beer than is needed to form a haze. The relationship of haze measurement by the EBC procedure, the ASBC method, and the NTU (nephelometric turbidity units) method as employed in some modern haze meters is shown in Table 23.

The major haze-forming polypeptides are derived from hordeins and are relatively rich in proline. Albumin- and globulin-derived polypeptides also come out of solution as chill haze, but only after the hordein-derived species.

There is a similarity between the polypeptide polyproline and the stabilizer polyvinylpolypyrollidone (PVPP), making it unsurprising that polyphenols recognize both PVPP and proteins containing regions rich in proline (Fig. 94). Hydrophobic amino acid residues may also be important, because it is through interactions between these that haze particles grow.

Haze-Active Polyphenols

Polyphenols in beer may arise from malt and hops. They are probably identical in their haze-potentiating capability.

Table 23. Comparison of haze measurements

Description	NTU[a]	EBC[b]	ASBC[c]
Brilliant	<2	<0.5	<35
Almost brilliant	2–4	0.5–1	35–70
Very slightly hazy	4–8	1–2	70–140
Slightly hazy	8–16	2–4	140–280
Hazy	>16	>4	>280

[a] Nephelometric turbidity units. 1 NTU = 0.25 EBC or 17.5 ASBC units.
[b] European Brewing Convention.
[c] American Society of Brewing Chemists.

Monomeric polyphenols such as catechin are incapable of forming a haze but do so readily once oxidized and converted into dimers and larger oligomers (tannoids). Oxidative polymerization occurs readily if beer is "punished" by heat (e.g., in a forced ageing regime), *even in the absence of oxygen.*

Tannoids and protein coexist in beer. It is claimed that there is a readily reversible equilibrium between the polyphenols and the proteins in their free and soluble but bound forms.

This is the model developed by L. Chapon:

$$P + T \leftrightarrow P - T \text{ (soluble)} \rightarrow P - T \text{ (insoluble)}$$

where P = haze-active protein and T = tannoid.

Its implication is that protein-tannoid complexes exist in beer in equilibrium with their building blocks and that this equilibrium can be shifted toward the building blocks by removing either protein or tannoid.

K. J. Siebert has described a more comprehensive model for protein-polyphenol interactions (Fig. 95). The model assumes that it is only proline-containing proteins that

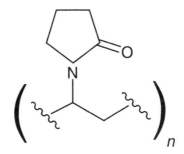

Fig. 94. Repeating unit in polyvinylpolypyrollidone (PVPP).

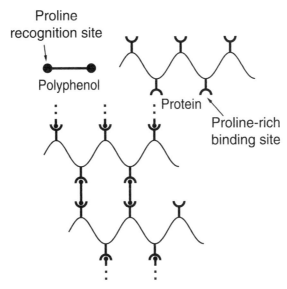

Fig. 95. Siebert model for haze formation.

interact to form chill haze, with a fixed number of polyphenol-binding sites in toto. Furthermore, it is assumed that a polyphenol has two (or more) "ends" that can specifically interact with these binding sites on proteins, thereby allowing a single polyphenol molecule to bridge between protein molecules. If there is an excess of haze-active protein over haze-active polyphenol (and this is usually so for beer), then most polyphenols are involved in bridging two proteins together, with insufficient polyphenol to bridge dimers and form larger particles. If the haze-active polyphenol is in excess of protein (for instance, as occurs in hard ciders), then there will be a shortage of free proline sites able to enter into cross-linking of protein molecules. It is only when there are roughly equal quantities of haze-active protein and haze-active polyphenol that the conditions exist for the formation of large networks that will manifest themselves as visible particles. Naturally, the levels of both need to be sufficiently high to manifest a visible haze when they associate.

Stabilization Treatments

Various stabilization treatments are available to lengthen the shelf life of a product. They should be applied as part of a more holistic approach to the achievement of colloidal robustness. There are three potential strategies:
- remove protein
- remove polyphenol
- remove a proportion of each

The binding forces involved in holding polymeric materials to adsorbents are relatively weak; hence, low temperatures are a prerequisite for colloidal stabilization.

Another key factor is beer clarity. Treatments involving PVPP and silica hydrogels are more efficient when applied to bright beer. Bright beer also has a lesser tendency to develop a haze through particle growth. If there are many small particles in beer (perhaps those responsible for invisible hazes), then they can form nucleation sites for haze development.

Brewery operations should be configured to minimize the load on the filter (e.g., possible use of kettle, isinglass, and auxiliary finings; efficient particle-settling regimes in cold conditioning, etc.). The filtration regime must be properly set up in respect of filter aid selection, dosing rate, lowest possible temperature, minimum oxygen pick-up, etc. Filter aid itself adsorbs haze materials.

PVPP

Two types of PVPP are presently in use for the removal of polyphenols:
- Single use, which comprises a micronized white powder of high surface-weight ratio. This readily adsorbs polyphenols on its surface and can therefore be incorporated into a filter aid body feed dosing regime.
- Regenerable, which can either be impregnated into sheets or used within devoted horizontal leaf pressure vessels. These are regenerated by hot 1–2% caustic and subsequent rinsing with hot and cold water and neutralization by CO_2 or 0.3% nitric acid. These materials are used only after powder-based filtration, and a trap filter of cotton or cellulose candles may be necessary after a powder-based system in order to retain very fine particles of PVPP.

Silica Hydrogels and Xerogels

Silicas adsorb haze-forming polypeptide. Silica hydrogels and xerogels are manufactured from sand. Xerogels are produced from hydrogels by drying before the milling stage that is used to derive the preferred particle sizes.

The critical features include the pore size of the particles and the surface area presented by them. The most effective pore size for a hydrogel is 30–120 Angstroms, and to use anything larger is to risk problems with reduced head retention. A reduction in particle size (giving an increase in surface area) increases the adsorption rate. This is of particular significance for the mode of use of hydrogels. If they are dosed into the storage tank, then time will allow equilibrium to be established. Conversely, if they are to be dosed in-line, then adsorption rate is an especially important parameter.

The level of silica hydrogel used depends on the intrinsic stability of the beer being treated, the shelf life required, and the extent to which other stabilizers are being used.

Silica hydrogel removes haze-forming protein but not foam polypeptide. This is because the silica hydrogel recognizes and interacts with the same sites on haze-active polypeptides as do the polyphenols.

Tannic Acid

Because it has a molecular structure closely similar to that of other polyphenolic species, tannic acid is claimed to be a relatively specific precipitant of haze-active proteins in beer (Fig. 96).

Tannic acid throws a sizeable precipitate when added to a cold-conditioning tank. This necessitates either cautious transfer of beer from sedimented material in tank or the use of a polisher centrifuge followed by membrane filtration. Alternatively, because the reaction

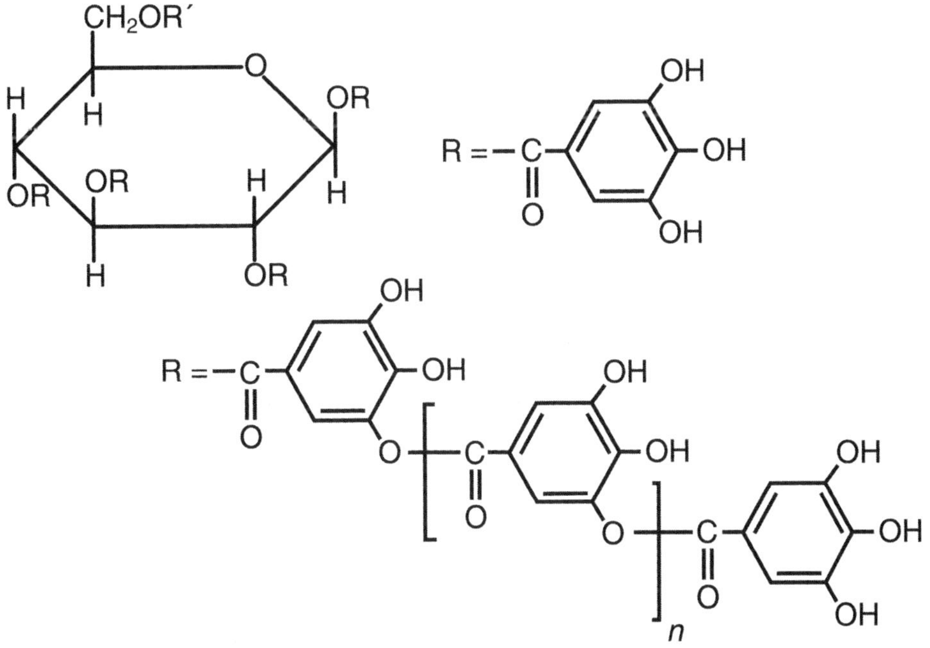

Fig. 96. Tannic acid.

time of new generation gallotannins is so rapid, they can be dosed on-line to the powder filter.

Enzymes

Papain, from *Carica papaya* (paw paw, papaya) was the first haze preventative employed in the brewing industry. The disadvantage of proteolytic enzymes as stabilizing agents is that they lessen foam quality. Recently, a proteinase that hydrolyzes peptide bonds involving a prolyl residue has come onto the market. Because of the importance of proline-rich sites in haze-forming proteins, it seems likely that this enzyme is more specific than papain.

Predictive Methods

Colloidal shelf life may be defined as the length of time before a beer displays a haze value of 2.5°EBC at 0°C. For many brewers, this (or a similar figure) is the yardstick against which they judge breakdown in a predictive test. Of these, there are two main types:
- hot/cold cycling tests
- precipitation tests

Hot/Cold Cycling

There are many variants of the hot/cold cycling test. Some do not even concern themselves with the cold stage; for instance, in the Guinness test, beer is held at 37°C. One week at this temperature is said to be the equivalent of one month of "normal" storage (18°C). Another example involves cycles of storage at 60°C for 2 days/–2°C for 1 day, in which one complete cycle is said to be the equivalent of 6 weeks of "normal" storage.

Such tests are useful in the study of trial brews, for example, the evaluation of new palliative procedures.

Precipitation Tests

The most famous precipitation test is the alcohol chilling test of Chapon. Alcohol is added as the temperature of a beer is lowered to –8°C, and chill haze is forced out within a total test time of 40 min. The test predicts only chill haze.

Another test—and one of those incorporated into the Tannometer (a commercial instrument that combines several predictive tests for haze life)—involves the precipitation of haze-active proteins by tannic acid, the so-called "protein sensitivity" of a beer. In a comparable test, gallotannin is replaced by a solution of saturated ammonium sulfate, hence its name the SASPL test (saturated ammonium sulfate precipitation limit). In the former, the amount of light scatter caused by a standard addition of tannic acid is measured. In the latter, the number of milliliters of $(NH_4)_2SO_4$ that need to be added to cause a measurable increase in turbidity is recorded—on the basis that the more salt is needed to bring out protein, the less such precipitable protein is in solution.

Tannoids can be quantified in a method analogous to the sensitive protein test by using polyvinylpyrrolidone monomer (PVP) as precipitant.

The Holistic Approach to Achieving Haze-Stable Beer

No brewer should expect to deal simply and efficiently with colloidal stability by application of one or more downstream stabilization techniques. The achievement of a colloidally robust beer depends on attention to detail from the selection of raw materials all the way through to packaging, storage, and distribution.

Raw Materials

There will be *pro rata* more potentially haze-forming protein in high-nitrogen batches.

Homogeneously well-modified malt precludes problems from β-glucans and allows efficient release and degradation of starch in mashing. Replacement of barley malt with syrup adjuncts, or with rice and maize grits or flakes, dilutes out all types of haze precursor. Barley- and wheat-based adjuncts, however, increase risks from haze-forming proteins, polyphenols, glucans, and, in the case of wheat, pentosans.

Barley varieties vary considerably in their content of polyphenols; for instance, winter varieties have relatively high levels. This is certainly not an argument against the use of such varieties, for it is the behavior of *specific* flavanoids that is important and not the absolute level of polyphenol.

The majority of the polyphenols are located in the husk. Alkaline steeping of barley removes polyphenol from the husk. Several low-polyphenol barley varieties have been developed.

Aroma hop varieties tend to contain higher levels of polyphenols. Various polyphenol-free bitterness preparations are available.

Brew House

Sweet wort production is probably best viewed as a protein-*precipitation* rather than a protein-*hydrolysis* stage. Low-temperature mashing-in, often called a "proteolytic stand," should be renamed a "β-glucanolytic stand," and it will certainly be necessary to deal with high-glucan grists unless exogenous β-glucanases are used.

Extraction of monomeric, dimeric, and trimeric flavanols is thought to be greater from single-temperature infusion mashes than from programmed mashes.

Oxygen directly or indirectly causes the oxidation of polyphenols, primarily through the action of peroxidases from malt. The effect is an increase in the polymerization index of the polyphenols, an attendant development of red color, and an increased precipitation due to the conversion of flavanol monomers and dimers to tannins.

The highest concentrations of all flavanols are found in the first runnings from a mash. However, more polymerized polyphenols are eluted maximally at the end of run-off. Perhaps the variation observed in the extent of polyphenol extraction during wort separation is a reflection of the varying extents to which oxidation has taken place during mashing and lautering. Prolonged run-off to extract weaker worts delivers proportionately more polyphenol into the wort if pH is allowed to drift too high (water pH of 6.5–7.0). The application of weak wort recycling can hardly be in keeping with attempts to maximize shelf life.

It is important that there be sufficient calcium in the grist to ensure precipitation of oxalate.

Of particular significance for colloidal stability is the kettle boil, with a well-formed hot break removed in the hop back or whirlpool. Apart from the levels of polyphenols and polypeptides, factors influencing the formation and removal of hot break include the extent of agitation and turbulence (a rolling boil affording enhanced opportunity for precipitation of materials at localized surfaces) and the use of coagulants such as carrageenan.

Fermentation and Maturation

Polyphenols may be removed by adsorption onto the yeast cell surface, and the extent to which this occurs depends on the freshness of the yeast and the extent of its growth.

Both from within and from its surface, yeast in poor condition (large number of generations, physical abuse such as that caused by centrifugation and adverse storage conditions, etc.) releases polysaccharide materials that can cause clarity problems.

The cold-conditioning stage is particularly critical. Beer should be chilled to as low a temperature as possible short of freezing (high-gravity beers can withstand lower temperatures, and thus beers produced using high-gravity brewing are potentially more stable than those brewed at sales gravity). The target for a beer of 5% (vol/vol) alcohol should be -1 or $-2°C$. Three days at this temperature is more effective than three weeks at $0°C$. A filter room should show evidence of frosted mains.

Judicious use of finings helps sedimentation of particles at this stage.

For many loss-conscious brewers, a primary consideration is the maximal recovery of beer from cold storage. Some blend back sediments on run-off from the cold tank, removing them by centrifugation. Others recover beer from tank bottoms in various ways, including by cross-flow filtration. There is an unavoidable risk that recovered beer will present a stability risk when blended back to the mainstream. Temperature and oxygen control are critical; otherwise, the dangers of returning haze precursors to beer are huge.

Brewers employing PGA as a foam-stabilizer should do so *strictly* according to manufacturers' instructions. It is probably worth paying slightly more for PGA already made into solution rather than to try to produce one's own PGA solutions.

Filtration and Packaging

Beer should be kept as cold as possible en route to and through the filter; otherwise, material precipitated during cold storage tends to return to solution and pass through into the bright beer.

Rigorous minimization of oxygen ingress is essential. Also essential is the avoidance of any release from filter aids of metal ions, such as iron and copper, as they will activate oxygen to the radical forms in which it is active. Precautions here include use only of additions with low metal loadings, avoidance of recycling on filters (especially during filter loading when high iron charges can be leached off filter aid), and use of water supplies with low metal content, perhaps achieved by treatment.

Filter aid selection is also critical to efficient removal of the dead bacteria that originate in malt, survive the brewing process, and enter into beer, where they cause dullness.

Attention should be paid to getting the gas contents within specification "right the first time" to avoid the need for gas washing with its risks of foaming and "skin" formation.

Cans should be properly screened to ensure that they do not have lubricant residues capable of generating haze.

Further Reading

Bamforth, C. W. (1999) Beer haze. J. Am. Soc. Brew. Chem. 57:81–90.

16. Flavor Instability

Achieving flavor stability is a bigger challenge than gaining haze stability, because the chemistry of flavor change is vastly more complex than is that of haze formation.

The chemical species responsible for flavor changes are often present in such low amounts that they can be difficult to detect, let alone measure accurately. By contrast, the haze-forming materials are relatively few in number, tend to be readily detectable, and should be easily removed.

The Nature of Flavor Changes Occurring in Beer

Changes in character caused by the formation or loss of materials in trace quantities (Fig. 97) are more perceptible in a bland beer than in one with fuller flavor.

Various changes may occur. Bitterness declines, and that remaining may become harsher. Fruity-estery and floral notes may decrease. A "ribes" (blackcurrant leaves, tomcat pee) character may become evident. Papery or cardboard characters will start to be detected. Progressively, beer will become bready and either sweeter or toffee-like and perhaps honey-like. Possibly metallic, earthy, or straw notes will develop. Finally, a beer will become woody and winey (sherry-like).

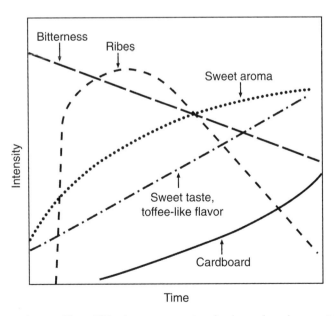

Fig. 97. Flavor changes in stored beer. This picture represents aging in an ale and may not fairly represent the changes occurring in a lager-style product.

As for all chemical reactions, the rate of flavor change depends on temperature, an increase in temperature of 10 degrees C leading to a two- to threefold increase in aging rate. Keeping beer cold is one of the most important means of maintaining freshness. When beer is held at 0–4°C, it fails to display signs of oxidation even after many months of storage. A beer packaged under best low-oxygen conditions might be expected to display signs of aging in around 100 days at 20°C.

This is why accelerated aging regimens employed in research studies of aging often use either 30°C for 30 days or 60°C for 1 day to mimic prolonged storage, although the precise nature of aging differs at different temperatures. A beer aged at 25°C tends to develop a predominantly caramel character, whereas at 30 or 37°C more cardboard notes are dominant.

Potential Changes in Flavor Substances During Beer Storage

Most of the substances deemed to be of importance for aged character, especially cardboard notes, contain the carbonyl group.

Addition of sufficient quantities of materials, such as sulfur dioxide (bisulfite), that can scavenge carbonyls will strip the oxidized flavor from beer in moments (Fig. 98).

Of all these carbonyls, (E)-2-nonenal is most frequently cited as the cause of a cardboard character in beer, but this is an oversimplification. It is likely that nonenal is just one of a related group of substances responsible for aged character.

The amounts of these materials are very low when compared to the concentrations of their purported precursors (next section). Consider nonenal, which has a flavor threshold of approximately 0.1 ppb. A substance will certainly be detectable in its own right if it is present at 5 flavor units. Assuming that nonenal originates from the oxidation of unsaturated fatty acids and supposing that there is 0.5 kg of linoleic acid per hectoliter of wort at 10°P, we can calculate that the efficiency of conversion of linoleic acid into nonenal need only be some $2 \times 10^{-5}\%$ for the required number of flavor units to be generated.

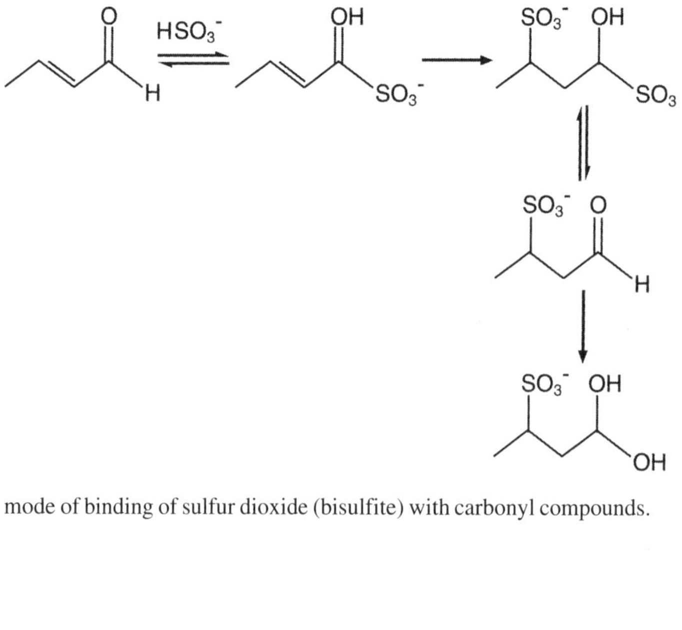

Fig. 98. The likely mode of binding of sulfur dioxide (bisulfite) with carbonyl compounds.

Pathways Involved in the Synthesis of Staling Substances

There is not universal agreement about which pathways are of especial importance for the origin of stale character in beer.

Melanoidin-Catalyzed Oxidation of Higher Alcohols

Alcohols in beer can be converted to their equivalent aldehydes through the catalysis of melanoidins.

Oxidation of Iso-α-Acids

Unhopped beers seldom develop oxidized flavor, indicating a possible role for iso-α-acids as precursors of staling compounds.

The reduced side-chain iso-α-acids used to impart light resistance to beers do not break down to staling carbonyls, and, as a result, beers bittered with such agents may be more resistant to oxidation.

Strecker Degradation of Amino Acids

The Strecker degradation involves a reaction between an amino acid and an α-dicarbonyl compound, such as the intermediates in browning reactions (Fig. 99). The amino acid is converted into an aldehyde with one fewer carbon atom. Polyphenols may have a catalytic role.

Aldol Condensations

The aldol condensation reaction between separate aldehydes or ketones is a plausible route through which (E)-2-nonenal may be produced by a reaction between acetaldehyde and heptanal (Fig. 100). Diverse other carbonyls may be generated in this way, with the imino acid proline (abundant in beer, see earlier) as a catalyst.

It is feasible that longer-chain staling aldehydes are built up in this way from the products of Strecker degradation, higher alcohol oxidation, or bitter substance breakdown.

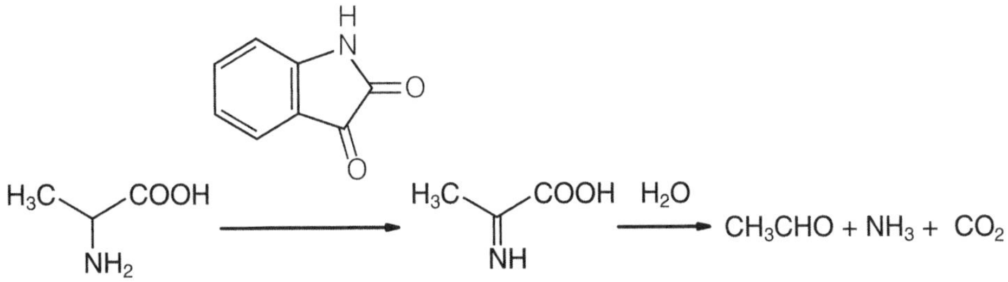

Fig. 99. Strecker degradation.

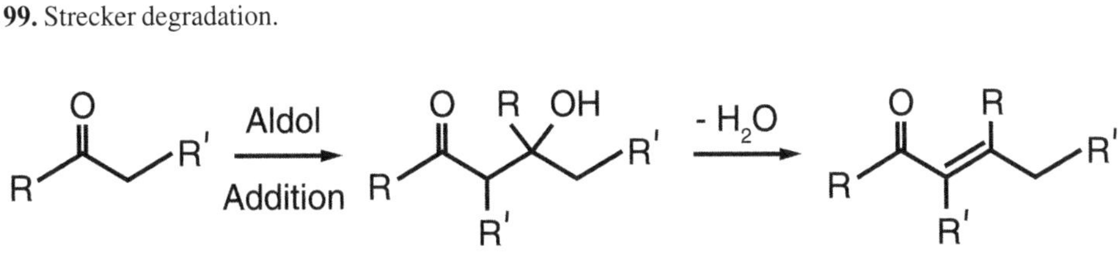

Fig. 100. Aldol condensation.

Oxidation of Unsaturated Fatty Acids

The oxidative breakdown of lipids leading to rancidity is a well-understood pathway in the aging of many foodstuffs. This may be through either or both of two routes: enzymic or nonenzymic oxidation.

Lipoxygenase-catalyzed oxidation of unsaturated fatty acids. Barley develops two lipoxygenase enzymes in its embryo, LOX-1 and LOX-2. Both enzymes are extremely heat-sensitive and are extensively lost during most kilning regimes, with LOX-1 being slightly the more resistant.

Both enzymes are capable of acting on polyunsaturated fatty acids (those fatty acids with two or more carbon-carbon double bonds). In effect, the only substrates of significance arising from malted barley are linoleic acid and linolenic acid, which have two and three double bonds, respectively, and the former is much more important.

LOX converts linoleic acid into hydroperoxy acids. Lipid oxidation will largely take place in the particles of a mash rather than in solution. In view of the heat sensitivity of lipoxygenase, it is likely that the enzyme is significant only at lower mashing temperatures (e.g., 52°C). Activity is also much less at lower pH, e.g., 5.0.

Nonenzymic oxidation of unsaturated fatty acids. The nonenzymic pathway leads to hydroperoxide species analogous to those emerging from the lipoxygenase route (Fig. 101).

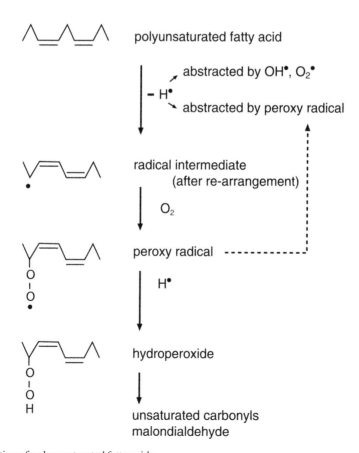

Fig. 101. Oxidation of polyunsaturated fatty acids.

Triggering this "autoxidation" process requires a highly active species able to abstract a hydrogen from the carbon atom adjacent to the double bond. The peroxy radical formed in the pathway is capable of effecting this abstraction, hence the autocatalytic nature of the process; however, only a limited number of radical species have this capability. Two of them are radical forms of oxygen, superoxide (in its protonated state, perhydroxyl) and hydroxyl (Figs. 102 and 103).

The production of radical forms of oxygen is promoted by light and certain enzyme systems. However, of particular importance are transition metal ions, for instance, iron and copper, hence the precautions that most brewers take to eliminate these substances from their process and product.

$HO_2^\bullet$ is much more reactive than O_2^-. As the pK_a for this acid-base equilibrium is 4.8, it will be realized that superoxide is likely to be a much more problematic species

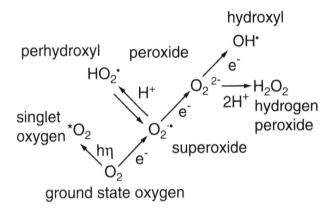

Fig. 102. Activation of oxygen. The electrons can be donated by metal ions such as iron and copper. $h\eta$ = light.

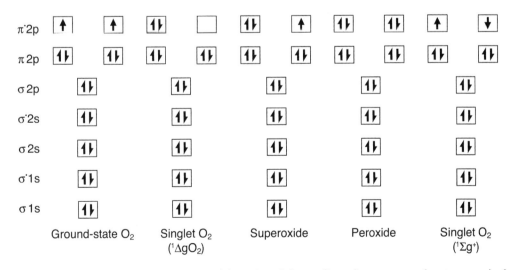

Fig. 103. Electron configuration in oxygen and its activated forms. Ground-state oxygen has two unpaired electrons in the outer orbital. It is with difficulty that it finds two electrons with opposite spin to "pair off," and so ground-sate oxygen is relatively unreactive. It is easier to find either one more electron or two with opposing spins. Hence, the other species shown are more reactive.

in beer than in wort. It has long been suggested that beers stale more quickly at lower pH.

The Central Role of Oxygen in Staling

It has long been recognized that oxygen levels in final pack should be as low as possible, and it remains true to this day that the top priority for any brewer wishing to have long shelf life in a product must be to package with the lowest practical oxygen levels. Modern filling equipment can achieve total oxygen levels in pack of less than 0.1 ppm.

Increasingly, though, it is accepted that even beers packaged under the most miserly of oxygen regimes still deteriorate in flavor. This has convinced people that oxygen control is needed throughout the brewing process and perhaps even in the malt house.

Oxygen in the brew house has several possible effects:
1. direct involvement in the lipoxygenase reaction
2. source of oxygen radicals for the autoxidation of unsaturated fatty acids
3. oxidation of sulfhydryl bonds in proteins, leading to cross-linking and retardation of lautering, but also to the production of hydrogen peroxide, which is a substrate for peroxidases (of which there are several very active ones in malt)
4. reaction (directly, or via H_2O_2 formation) with polyphenols, catalyzed by peroxidases in the mash but nonenzymically in the wort boil.

The last of these effects may be of especial significance, for the polyphenols are the prime antioxidant materials native to malt, hops, and, subsequently, beer. If they are lost by oxidation at this stage (they polymerize on oxidation and precipitate with protein), then the subsequent beer will have less antioxidant capacity.

Endogenous Antioxidant Systems

Polyphenols. Owades used $^{18}O_2$ to show that when oxygen is introduced into beer, 5% enters iso-α-acids, 35% enters volatile carbonyls, but the greater part goes into the polyphenol fraction.

Polyphenols may be antioxidants in brewing systems for at least three reasons:
1. their ability to scavenge oxygen free radicals, superoxide and hydroxyl, and also the peroxy radicals formed in the autocatalytic oxidation of unsaturated fatty acids
2. their capability as inhibitors of lipoxygenase
3. their capacity as chelating agents, which sequester transition metal ions such as iron and copper.

Melanoidins. Melanoidins, too, are capable of scavenging superoxide, peroxide, and hydroxyl, and it is claimed that they suppress the oxidation of iso-α-acids and unsaturated fatty acids. However, as we have seen, melanoidins are also implicated in *promoting* the oxidation of higher alcohols.

Sulfur dioxide. SO_2 is also a radical scavenger, and it has been claimed that this is its main antioxidative function in beer. However SO_2 can also form adducts with carbonyl compounds, rendering complexes that are not flavor-active. The proposal is that carbonyl compounds produced early in the brewing process bind with the SO_2 that is a natural product of yeast metabolism. These adducts pass into beer, where they progressively break down to free the carbonyls, which render the beer stale. The bisulfite complexes are in equilibrium with the free carbonyl and SO_2. Thus, by maintaining as high as possible a free-SO_2 level in beer, perhaps by adding metabisulfite (within legal constraints), the equilibrium is maintained in favor of the binding of carbonyl.

The problem is that SO_2 is progressively lost from beer in reaction(s) that have not yet been identified (but may include elimination by reaction with oxygen), thereby shifting the balance in favor of carbonyl release. Loss of SO_2 is brand sensitive and, in particular, is promoted by increased temperature (another clear justification for keeping beer cold): at 40°C, half of the SO_2 in beer is lost in 27 days, whereas the half life is approximately 3 years at 0°C.

The only endogenous source of SO_2 in beer is through production by yeast, from the reduction of sulfate in water and grist materials. Factors influencing the extent of SO_2 production by yeast will be reviewed later.

It has been suggested that amino groups within proteins also bind carbonyl compounds in an analogous manner to SO_2.

Yeast. Perhaps the most powerful "reducing agent" involved in brewing is yeast. Apart from producing sulfur dioxide, it is also capable of reducing carbonyl compounds (*apart* from those in adduct form with SO_2) to their equivalent alcohols.

Several enzymes in yeast seem to be involved in the reduction of carbonyl compounds, including the alcohol dehydrogenase that is involved in ethanol production.

Chelation. Apart from the polyphenols, a myriad of other species in wort and beer have a strong ability to chelate metal ions, notable amongst them amino acids, phytic acid, and, of course, melanoidins. Transition metal ions bound with chelators sometimes have less capability to promote oxygen radical formation, but sometimes more. It is wisest to minimize the opportunity for ions such as copper and iron to gain access to the product at any stage in the process.

The Brewing Process in Relation to Flavor Stability

Grist. In malt there are
- (possibly) preformed staling precursors
- substrates for staling reactions
- enzymes capable of promoting flavor deterioration (lipoxygenases)
- enzymes capable of scavenging oxygen and oxygen radicals (peroxidases and superoxide dismutases)
- other substances that can react with oxygen (e.g., sulfhydryl groups in proteins)
- various antioxidants, including polyphenols, phenolic acids, and melanoidins.

Assuming that lipoxygenase activity is unwanted, then it is clear that malts kilned to relatively intense regimes will contain pro rata less lipoxygenase.

Any technique that reduces embryo growth during malting would be to the benefit of flavor stability: for example, use of potassium bromate or high hydrostatic pressures.

More-highly-kilned malts (especially roasted malts) have increased antioxidant levels.

Brew house. Considerable quantities of oxygen may be consumed in the brew house. Although unproven, this may impact flavor life, and it certainly affects color formation and rates of wort separation.

Various straightforward precautions can be taken to lessen the ingress of air in the brew house.
- use of a premasher (cf., Steel's masher) to enable intimate mixing of grist and water under low-air conditions
- filling of vessels by bottom entry
- not switching on agitators until they are covered, and then only using them where absolutely necessary to ensure equilibration of vessel contents and temperature
- good housekeeping in respect to pipe and pump integrity
- switching off pumps as soon as vessel contents have been transferred, to avoid the risk of sucking air into the system
- use of nitrogen as a motor gas.

Assuming that lipoxygenase is disadvantageous for flavor stability, then its action will be reduced by lowering the level of one of its substrates, oxygen, but also by mashing-in at the highest temperature commensurate with other mashing requirements.

Gentler milling will leave the embryo intact and therefore mean less extraction of LOX.

Lowering the mashing pH to 5.1 reduces lipoxygenase action.

Copper vessels will provide a sizeable charge of ions capable of activating oxygen to its damaging radical forms.

In the kettle boil, there is the opportunity for driving off preformed staling substances.

Thermal degradation reactions at the whirlpool stage may generate undesirable flavor compounds, and it has been suggested that application of a vacuum (0.5 bar) after this stage is advantageous.

Some advocate the use of a trub separator to enhance shelf life.

Fermentation. Oxygen should be introduced on the cold side of the heat exchanger, and only as much as is needed by the strain in use. Direct oxygenation of yeast precludes any opportunity for wort oxidation.

Yeast has at least two direct impacts on flavor life: its ability to produce sulfur dioxide and its ability to reduce carbonyl compounds.

It is important to ensure the vigor of a fermentation by using established procedures for having yeast of high viability and vitality, pitched in the appropriate quantities.

SO_2 levels are increased if the sulfate supply to the yeast is increased, wort clarity is increased, wort oxygenation and pitching rate are lowered, and fermentation temperature is reduced. Yeast strains capable of increased sulfite production have been developed.

Downstream processing and packaging. Downstream of the fermenter, the overriding priority is to minimize oxygen pick-up. This is achieved through
- use of nitrogen or carbon dioxide as a motor gas
- blanketing vessels with either of these gases prior to filling
- rigorous deaeration of water (to less than 30 ppb O_2) used for any purpose (other than cleaning) downstream (e.g., dilution, slurrying of filter aids, make up of additions)
- precautions with pumps, pumping, and leaks.

A realistic target for bright beer prior to packaging should be 50 ppb.

Cold beer reacts only slowly with oxygen, and therefore any high O_2 levels can be corrected at this stage (it is not possible to rectify O_2 ingress into hot streams in the same way) by sparging with inert gas or, preferably, using hydrophobic membrane technology.

At the packaging stage it is crucial that
- filler bowls are counter-pressured with inert gas
- the bowl is continuously purged with the gas
- containers are pre-evacuated, preferably twice, separated by a CO_2 purge
- beer is fobbed (foamed) prior to capping (bottles) or seaming (cans), e.g., through use of high-pressure water jetting for bottles or undercover gassing for cans.

Exogenous antioxidants and protectants. Various antioxidants may be used in beer (depending on regional legislation), with the most prominent of these being sulfur dioxide and ascorbic acid. SO_2 is much the more effective, either through its role as a carbonyl binder or as a radical scavenger.

These agents can protect only against *new* oxidation occurring in the beer; they will not rectify any damage that has occurred upstream. They will help protect against oxygen leaking into bottled beer through the crown cork-bottle seal, which is now established as a serious problem. Oxygen-barrier and oxygen-scavenging bottle closures are available.

The advent of plastics with enhanced barrier properties has led to some mainstream brands being filled into such lightweight bottles.

Predictive Tests for Aging

Accelerated aging regimes were mentioned earlier.

Electron spin resonance (ESR) procedures that detect free radicals have recently been championed as means for predicting staling.

Further Reading

Bamforth, C. W. (2004) Fresh controversy: Conflicting opinions on beer staling. Proc. Conv. IGB Asia-Pacific, pp. 63-73.

17. Foam

The Physics of Foaming

The key processes leading to the collapse of beer foam are drainage and disproportionation. The most stable foams are those with a homogeneous distribution of small bubbles. If the size population is mixed, then gas will pass from small bubbles to adjacent larger bubbles, in the process known as disproportionation, making the small bubble even smaller until it disappears and the large bubble is even larger. The consequence is fewer bubbles (i.e., less foam) and bigger, bladdery bubbles that are less attractive.

The equation describing disproportionation is from De Vries:

$$r_o^2 - r_t^2 = \frac{4RTDS\sigma\, t}{P_{atm}\, \theta}$$

where r_o = bubble radius at t = 0
 r_t = bubble radius at t = t
 t = time
 R = universal gas constant
 T = absolute temperature
 D = diffusion coefficient
 S = solubility of gas
 σ = surface tension
 P_{atm} = atmospheric pressure
 θ = film thickness between bubbles

In foams of uniform bubble size, surface collapse is more important. The speed of collapse is proportional to the rate at which successive layers of bubbles come to the surface. Thus, if the bubbles in a foam are small, there are more layers of bubbles in a given depth of foam. Liquid draining from a foam of small bubbles also has a more convoluted route to take (because of the greater surface area), and this is another reason why foams composed of smaller bubbles are more stable.

The equation governing drainage is

$$Q = \frac{2\rho g q \delta}{3\eta}$$

where Q = flow rate (m^3 s^{-1})
 η = viscosity of film liquid

ρ = density
q = length of plateau border (m)
g = acceleration due to gravity
δ = thickness of film (m)

Beer Polypeptides and Foam

There are two schools of thought:
1. There are certain proteins in beer that make a primary contribution to foam stability—the discrete polypeptide hypothesis.
2. The essential feature of a protein or polypeptide determining its foamability is its hydrophobicity. Individual polypeptides in beer may exist in different conformational states, and it is when they present a hydrophobic exterior that they are particularly surface active—the generalized amphipathic polypeptide hypothesis.

Lipid transfer protein (LTP1) is a barley-derived protein and a significant component of beer foam. Neither malting nor mashing has any effect on the level entering wort. LTP1 is transformed (by denaturation) during wort boiling into a much more foam-stabilizing form.

A protein with a molecular weight of 40,000 known as Protein Z may also be important. (It is the protein that binds to β-amylase in barley.)

The alternate hypothesis is that it is hydrophobicity that is the primary determinant of foam-stabilizing ability in beer polypeptides: polypeptides with increased hydrophobicity display higher foam-stabilizing capability. Both LTP1 (after boiling) and Protein Z have hydrophobicity—but so do other polypeptides.

At least some of these polypeptides originate from the barley storage protein (hordein). Recent evidence suggests that these polypeptides have superior foamability but lower foam stability than the albumin-derived polypeptides (LTP1, protein Z). They outcompete LTP1 and protein Z to get into the foam, but as a result, the foam is less good than if LTP1 and Protein Z preferentially entered the foam.

Lipid-binding proteins, found especially in wheat (puroindolenes) but also in barley, enhance foam by binding inhibitory lipid. The enhancement of beer foam by the use of isinglass finings probably reflects their lipid-binding capability.

Foaming polypeptides are lost by precipitation throughout brewing, but particularly during wort boiling, in the yeast head on fermentation, and during filtration.

Other Components of Beer Foam

Melanoidins are capable of stabilizing foams.

When present alone as foaming species, both proteins and surfactants (such as lipids and detergents) can give good foams, whereas when both are together simultaneously, the foam is of inferior quality because the different types of molecule stabilize foams in different ways (Figs. 104–106).

The iso-α-acids cross-link polypeptides via hydrophobic and ion-dipole interactions (Appendix 1). The iso-α-acids probably interact through hydrophobic bonds with adjacent hydrophobic polypeptides, thereby "locking up" the foam surface. Divalent cations promote the interaction. The *trans* isomers of the individual iso-α-acids

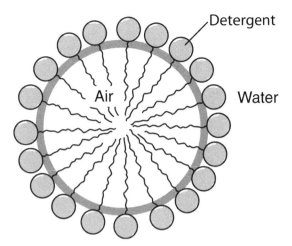

Fig. 104. Detergent-stabilized bubble.

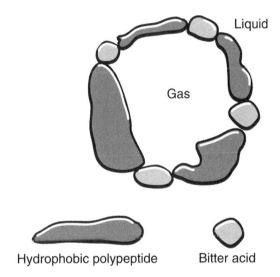

Fig. 105. Protein-iso-α-acid stabilized foam.

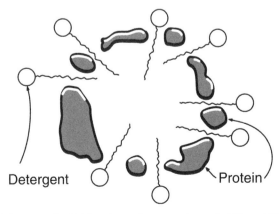

Fig. 106. Presence of both detergent (or lipid) and protein means that no bubble stabilization is achieved.

concentrate in beer foam to a greater extent than the corresponding *cis* isomers, with especial concentration of the *trans* isomers of isohumulone and isoadhumulone.

The reduced iso-α-acids are substantially more foam active than their unreduced counterparts due to their increased hydrophobicity.

It has been suggested that certain amino acids (the basic ones, arginine > lysine > histidine) interfere with the interaction of polypeptides and iso-α-acids, thereby inhibiting lacing.

Ethanol inhibits foam, probably by acting in a fashion analogous to that of lipids. At low concentrations, ethanol promotes foam formation by lowering surface tension.

Process Effects on Foam

Barley. Higher N levels will lead to higher levels of foam protein.

Malting. Hydrolysis of hordein releases foam-active polypeptides. Protein Z is released from linkage to β-amylase.

Wort production. Beer produced under high-temperature mashing conditions has a better foam. Acidification of the mash (to pH 5.1) also benefits foam, as does the use of malt containing high levels of melanoidins (e.g., crystal malt).

Fermentation and maturation. High-gravity brewing results in beers of inferior foam performance. This is due to such beers containing a lower quantity (at least by 40%) of hydrophobic polypeptide, which is lost in increased proportions during boiling and fermentation. It also seems that less of this foam-active polypeptide is extracted during the mashing of high-gravity worts. Yields of hydrophobic polypeptide were higher through a mash filter than a lauter tun.

Yeast proteinase A reduces the foam quality of beer by hydrolyzing the hydrophobic polypeptide fraction. Yeast of low vitality yields high protease levels and correspondingly poor beer foams. Yeast should be separated from the beer as soon as is practical. A slow cooling after fermentation yields a better beer foam than a fast cooling. Proteinase A release depends on the yeast strain (top-fermenting strains apparently produce more) and is increased when yeast is under stress (e.g., nitrogen starvation, high alcohol levels, high CO_2, high pressure).

Pasteurization destroys the proteinase, but the enzyme survives in sterile-filtered beer, which means that foam quality progressively deteriorates in the latter.

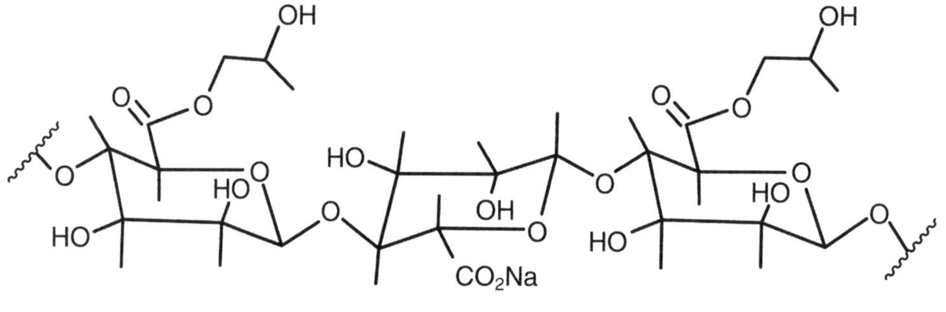

Fig. 107. Propylene glycol alginate (PGA).

Novel Solutions for the Enhancement of Beer Foam

Nitrogen gas enhances foam stability. For the re-creation of draught-style beers in small pack, brewers have linked the use of nitrogen to the widget (a plastic or metal device in cans or bottles that promotes nucleation). Alternatively, etched drinking glasses may be used to promote the replenishment of foam for any beer style.

Properly configured and clean dispense conditions with respect to pumps, pipes, taps, and glasses are a key factor determining foam quality. It is only to a limited extent that protectants against lipid damage such as PGA (Fig. 107) will be able to overcome deficiencies in the trade or the home.

Assessment of Beer Foam

For QC/QA purposes, the most frequently used methods worldwide are the procedures of Rudin (Fig. 108), Ross and Clark (ASBC standard), and NIBEM (Fig. 109).

Decisions on whether to release beer to trade are most rapidly and effectively made by little more than a shaking test. A fixed volume of finished beer is shaken in a closed cylinder and allowed to stand for a few minutes. Any beers with severe foam problems will be evident from the appearance of the head. Such a test should be applied to representative batches from all bright beer tanks and packaged beers.

Methods for the assessment of cling include the lacing index technique and another procedure from NIBEM.

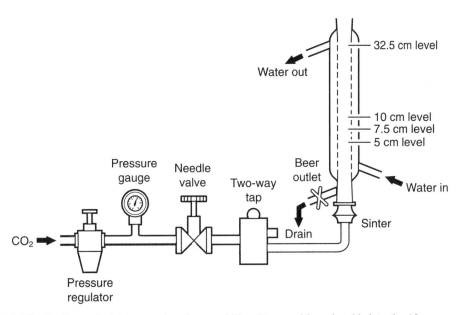

Fig. 108. The Rudin method for assessing foam stability. Degassed beer is added to the 10-cm mark and foamed by injecting carbon dioxide (or nitrogen) at a controlled rate and constant temperature until the foam reaches the 32.5-cm level. The gas flow is then stopped. In this way, the whole column is filled with foam. When the foam-liquid interface reaches the 5-cm mark, a stopwatch is started, and the time taken for the interface to reach the 7.5-cm mark is recorded. The longer this time, the slower the rate of liquid drainage and the more stable the foam.

166 / Chapter 17

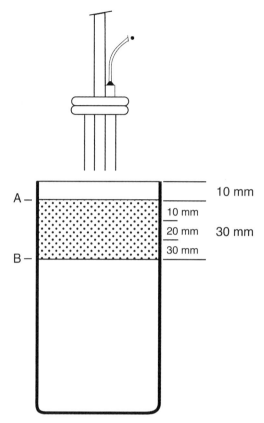

Fig. 109. The NIBEM method for assessing foam stability. Beer is introduced to the glass such as to produce a constant foam. The needles measure conductivity. When the foam collapses and contact with the needles is lost, the machine responds by lowering the needles until conductivity is once more established. The rate at which the needles are lowered is in proportion to the stability of the foam.

Further Reading

Bamforth, C. W. (2004) The relative significance of physics and chemistry for beer foam excellence: Theory and practice. J. Inst. Brew. 110:259-266.

18. Gushing

Gushing is the spontaneous generation of foam on opening a package of beer. When the excess pressure on top of a bottle or can of beer is released, many fine bubbles form throughout the beer and ascend very rapidly.

Type I gushing is caused by solid particles in beer, which act as nucleation sites for bubble release.

Type II gushing is due to stable microbubbles of gas, produced by agitation of a beer and its equilibration to room temperature. Pasteurization may temporarily eliminate it. It is understood that microbubbles form by foaming and trap CO_2 within extremely stable and impervious walls.

Gushing is also classified into "primary" and "secondary" gushing.

Primary gushing is caused by infection of the raw materials used to make beer, notably barley. It manifests itself as outbreaks of gushing. The infection is principally by the mold *Fusarium*, although other molds (*Rhizopus, Aspergillus, Alternaria,* and *Penicillium*) are sometimes responsible. The problem can be overcome by steeping barley in formaldehyde, but modern malting operations wouldn't use such a material.

The presence of *Fusarium* is in itself not a good predictor of gushing, although it is clearly inadvisable to malt barley that is infected. Even if gushing does not develop, such organisms can produce mycotoxins such as deoxynivalenol (DON; vomitoxin, see earlier).

Mold growth will occur particularly when barley encounters damp conditions during growing; hence, gushing can be a prime problem with barley grown in the far north of Europe. However, gushing is also induced in other regions in years of extreme drought.

It is believed that the fungi responsible for gushing produce a small (approximately 15,000 mol wt) polypeptide (hydrophobin) of extreme hydrophobicity and capable of causing gushing when present at 50 ppb.

Secondary gushing is caused by a range of agents that are capable of nucleating microbubbles:

- calcium oxalate crystals
- slivers of glass in new glass bottles that have been inadequately prewashed
- rough surfaces on the inside of glass bottles
- heavy metals such as nickel
- oxidized and dimerized resins in old hops and in bittering extracts
- filter aid breakthrough
- haze particles
- excess carbonation

If most bottles gush, this indicates primary gushing. If there is variability between cans or bottles, then this indicates some form of secondary gushing.

Several gushing tests have been developed, but none has yet been fully accepted. They all involve some degree of agitation and have been developed to allow screening of beers made on a small scale from worts from suspect barley or of extracts from malt or hop products for their tendency to promote gushing when added to packaged beer.

Further Reading

Casey, G. P. (1996) Primary versus secondary gushing and assay procedures used to assess malt/beer gushing potential. Tech. Q. Master Brew. Assoc. Am. 33:229–235.

19. Light Instability

Exposure of beer to light causes it to develop a "skunky" aroma. This is referred to as "light-struck" or sometimes "sun struck."

It is primarily due to a substance called 3-methyl-2-butene-1-thiol (MBT), although there are claims that methanethiol and methional also contribute.

Tasters differ in their sensitivity, but the range of flavor thresholds is 0.4–35 ng/liter.

MBT is formed from the isopentenyl side-chain of the bittering compounds (iso-α-acids) in a light-triggered reaction. Wavelengths in the range 350–500 nm are most damaging. Riboflavin potentiates the reaction. The S is provided by either a low molecular weight source (such as H_2S) or a high molecular weight source (viz., a polypeptide).

Brown glass provides far better protection than green or clear ("flint") glass, because, for the most part, it does not allow the passage of the key wavelengths of light (Fig. 110). However, marketing pressures have forced many brewers to package beer into green or clear glass bottles, the argument being that these are more attractive. In such cases, the beer should be kept out of light as much as possible (e.g., in cardboard boxes) and certainly should not be left on illuminated shelves in bars; only a few minutes are needed for MBT to be formed.

If the iso-α-acids are reduced, then they will no longer yield MBT. Iso-α-acids are extracted from hops using liquid carbon dioxide and then chemically reduced (Fig. 111).

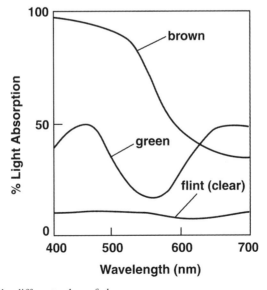

Fig. 110. Light absorption by different colors of glass.

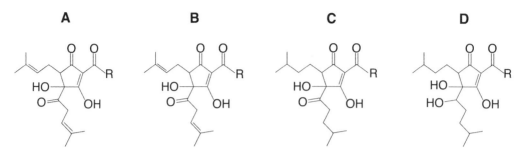

Fig. 111. Reduced iso-α-acids. A, Non-reduced; B, rho-; C, tetra-; D, hexa-.

Successively, one obtains rho-iso-α-acids (2 hydrogens picked up), tetrahydro-iso-α-acids (4H), and hexahydro-iso-α-acids (6H). They are progressively more resistant to light damage. They are also more bitter than the unreduced iso-α-acids and therefore need to be used in lesser amounts to yield the same perceived bitterness.

It is critical that there should be absolutely no unreduced iso-α-acid in a beer to achieve total resistance to a light-struck reaction. (Because the flavor threshold for MBT is so low, you don't need much iso-α-acid breakdown to go above flavor threshold.) Brewers should remember that iso-α-acids do accumulate on yeast, and so no yeast that has been used to ferment a "normally hopped" beer should be used to pitch a fermentation destined for 100% bittering by reduced side-chain iso-α-acids.

Further Reading

Templar, J., Arrigan, K., and Simpson, W. J. (1995) Formation, measurement and significance of lightstruck flavor in beer. Brew. Dig. 70(5):18–25.

20. Biological Instability

Compared with most other foods and beverages, beer is relatively resistant to infection. There are several reasons for this:
- presence of ethanol, which has some antimicrobial properties
- low pH
- relative shortage of nutrients (sugars, amino acids)
- anaerobic
- presence of antimicrobial agents (possibly the polyphenols and certainly the iso-α-acids)

Bacteria

The most problematic Gram-positive bacteria are lactic acid bacteria belonging to the genera *Lactobacillus* and *Pediococcus*.

Some are heterofermentative (producing lactic acid and other acids or alcohols and perhaps diacetyl) or homofermentative (producing only lactic acid). They tolerate acid. Some species (e.g., *L. brevis* and *L. plantarum*) grow quickly during fermentation, conditioning, and storage, whereas others (e.g., *L. lindner*) grow relatively slowly. Spoilage with lactobacilli is especially problematic during the conditioning of beer and after packaging, resulting in a silky turbidity and off-flavors.

Pediococci are homofermentative. Six species have been identified, the most important being *P. damnosus*. Such infection generates lactic acid and diacetyl. The production of polysaccharide capsules can cause ropiness in beer. The beer appears to be oily when poured due to the movement of the slimy polysaccharide.

Many Gram-positive bacteria are killed by iso-α-acids. These agents probably disrupt nutrient transport across the membrane of the bacteria, but only when they are present in their protonated forms (i.e., at low pH). This is one of the reasons why a beer at pH 4.0 will be more resistant to infection than one at pH 4.5. Some Gram-positives are resistant to iso-α-acids and most Gram-negatives are.

Important Gram-negative bacteria include the acetic acid bacteria (*Acetobacter*, *Gluconobacter*); Enterobacteriaceae (*Escherichia*, *Aerobacter*, *Klebsiella*, *Citrobacter*, *Obesumbacterium*); *Zymomonas*, *Pectinatus*, and *Megasphaera*.

Acetic acid bacteria produce a vinegary flavor to beer and a ropy slime. They are most often found in draft beer, where there is a relatively aerobic environment close to the beer, e.g., in partly emptied containers.

Enterobacteriaceae are aerobic and can't grow in the presence of ethanol. They are a threat in wort and early in fermentation, and they produce cabbagy-vegetable-eggy aromas. Heat-tolerant *Enterobacter* infecting the brew house may produce nitroso compounds (which pose a health risk) through their ability to reduce nitrate.

Zymomonas is a problem with primed beers (it uses invert sugar or glucose but can't use maltose). Although it has a metabolism reminiscent of *Saccharomyces* (it's actually used to produce alcoholic beverages in some countries), it does tend to produce large amounts of acetaldehyde.

Pectinatus bacteria are strictly anaerobic and are rod-shaped. They can spoil packaged beer, producing acetic and propionic acids, acetoin, and hydrogen sulfide.

Megasphaera bacteria are strictly anaerobic cocci capable of generating butyric acid and certain other acids. This genus is likeliest a problem in lower alcohol beers (<3.5% ABV).

Wild Yeast

A wild yeast is any yeast other than the culture yeast used for a given beer. As well as *Saccharomyces,* wild yeast may be *Brettanomyces, Candida, Debaromyces, Hansenula, Kloeckera, Pichia, Rhodotorula, Torulaspora,* or *Zygosaccharomyces.*

If the contaminating yeast is another brewing yeast, then the risk is a shift in performance to that associated with the "foreign" yeast (i.e., you won't get the expected beer). If the contaminant is another type of yeast, the risk is off-flavor production (e.g., clove-like flavors produced by decarboxylation of ferulic acid) or a problem such as over-attenuation, as might be caused by a diastatic organism such as *S. diastaticus.*

Further Reading

Priest, F. G., and Campbell, I., eds. (2003) Brewing Microbiology, 3rd ed. Kluwer Academic/Plenum Publishers, New York.

21. Packaging

The packaging operation is the most expensive stage in the brewery, in terms of raw materials and labor.

Beer will be brought into specification in the bright beer tank (sometimes called the fine ale tank or the package release tank). The carbonation level may be higher (e.g., 0.2 vol) than that specified for the beer in package to allow for losses during filling.

Removing Microorganisms

Although beer is relatively resistant to spoilage, it is by no means entirely incapable of supporting the growth of microorganisms. For this reason most beers are treated to eliminate any residual brewing yeast or infecting wild yeasts and bacteria before or during packaging. This can be achieved in one of two ways: pasteurization or filtration.

Pasteurization

Pasteurization can take one of two forms in the brewery: flash pasteurization for beer pre-package and tunnel pasteurization for beer in can or bottle. The principle in either case is that heat kills microorganisms.

One pasteurization unit (PU) is defined as 1 min at 60°C.

The higher the temperature, the more rapidly are microorganisms destroyed. A rise in temperature of 7 degrees C leads to a tenfold increase in the rate of cell death. The pasteurization time required to kill organisms at different temperatures can be read from a plot. Typically a brewer might use 5–20 PU, but higher "doses" may be used for some beers, e.g., low-alcohol beers, which are more susceptible to infection.

Good brewers will ensure low loadings of microorganisms by attention to hygiene throughout the process and by ensuring that the prior filtration operation is efficient.

In flash pasteurization, the beer flows through a heat exchanger (essentially like a wort cooler acting in reverse) that raises the temperature, typically to 72°C. Residence times of between 30 and 60 sec at this temperature are sufficient to kill off virtually all microbes. The configuration of the flash pasteurizer is such that heat from the beer leaving the device is used to warm that entering. It is essential that the oxygen level of the beer be as low as possible (preferably below 0.1 ppm) before pasteurization, because when temperatures are high, oxygen is "cooked" into the product, giving unpleasant flavors.

Tunnel pasteurizers comprise large heated chambers through which cans or glass bottles are conveyed over a period of minutes, as opposed to the seconds employed in a flash pasteurizer. Accordingly, temperatures in a tunnel pasteurizer are lower, typically 60°C for a residence time of 10–20 min.

Sterile Filtration

The rationale for selecting this procedure rather than pasteurization is as much for marketing reasons as for any technical advantage it presents. Many brands of beer these days are being sold on a claim of not being heat-treated and therefore being free from any "cooking." In fact, provided the oxygen level is very low, modest heating of beer does not have a major impact on the flavor of many beers, although those products with relatively subtle, lighter flavor will obviously display "cooked" notes more readily than will beers that have a more complex flavor character.

The sterile filter must be located downstream from the filter that is used to separate solids from the beer. Sterile filters may be of several types; a common variant incorporates a membrane formed from polypropylene or polytetrafluoroethylene and contains pores of between 0.45 and 0.8 μm.

Filling Operations

Filling Bottles

Bottles entering the brewery's packaging hall are first washed, irrespective of whether they are one-trip or returnable. The former will receive simply a water wash, as the supplier will have been required to make sure they arrive at the plant in a clean state. Returnable bottles, after they have been automatically removed from their crate and delivered to conveyors, need a much more robust cleaning and sterilization, inside and out, involving soaking and jetting with hot caustic detergent and thorough rinsing with water. Old labels will be soaked off in the process. En route to the filler, the cleaned and sterilized bottles pass an empty bottle inspector (EBI), a light-based detection system that spots any foreign body lurking in the bottle.

The beer coming from the bright beer tanks (i.e., after filtration) is transferred to a bowl at the heart of the filling machine. Bottle fillers are machines based on a rotary carousel principle. They have a series of filling heads: the more heads, the greater the capacity of the filler. Modern bottling halls are capable of filling in excess of 1,200 bottles per minute.

The bottles enter on a conveyor and, sequentially, each is raised into position beneath the next vacant filler head, each of which comprises a filler tube. An airtight seal is made and, in modern fillers, a specific air-evacuation stage starts the filling sequence. The bottle is counter-pressured with carbon dioxide before the beer is allowed to flow into the bottle by gravity from the bowl. The machine will have been adjusted so that the correct volume of beer is introduced into the vessel. Once filled, the "top" pressure on the bottle is relieved, and the bottle is released from its filling head. It passes rapidly to the machine that crimps on the crown cork. En route, either the bottle will have been tapped or its contents "jetted" with a minuscule amount of sterile water in order to fob the contents of the bottle and drive off any air from the space in the bottle between the surface of the beer and the neck (the "headspace").

The next stop is the tunnel pasteurizer (see earlier) if the beer is to be pasteurized after filling; although as we have seen, more and more beer is sterile filtered and packaged into already sterilized bottles. In the latter case, the filler and capper will tend to be enclosed in a sterile room into which only necessary personnel are allowed.

The bottles now pass via a scanner, which checks that they are filled to the correct level, to the labeler, where labels are rolled onto the bottles, and then perhaps to a device that applies foil to the neck. Other specialist equipment may involve jetting on a packaging date or "best before" date. Finally the bottles are picked up by a machine that places them carefully into a crate, or box, or whichever is the secondary package in which they will be transferred to the customer. Perhaps they will go straight from this operation onto a truck or rail car for shipping, but more often they will be stored carefully in a warehouse prior to release.

Canning

Cans may be of aluminum or stainless steel, which will have an internal lacquer to protect the beer from the metal surface and vice-versa. They arrive preprinted in the canning hall on trays. They are inverted, washed, and sprayed prior to being filled in a manner very similar to bottles. Once filled, the lid is fitted to the can basically by folding the two pieces of metal together to make a secure seam past which neither beer nor gas can pass.

Kegging

Kegs are manufactured from either aluminum or stainless steel. They are containers generally of 1 hl or less, containing a central spear. Kegs, of course, are multi-trip devices. On return to the brewery from an "outlet," they are washed externally before transfer to the multi-head machine in which successive heads are responsible for their washing, sterilizing, and filling (with beer that is either sterile-filtered or flash pasteurized). Generally they will be inverted as this takes place. The cleaning involves high-pressure spraying of the entire internal surface of the vessel with water at approximately 70°C. After about 10 sec, the keg passes to the steaming stage, the temperature reaching 105°C over a period of perhaps half a minute. Then the keg goes to the filling head, where a brief purge with carbon dioxide precedes the introduction of beer, which may take a couple of minutes. The discharged keg is weighed to ensure that it contains the correct quantity of beer and is labeled and palleted before warehousing.

Further Reading

Broderick, H. M., ed. (1982) Beer Packaging. Master Brewers Association of the Americas, Madison, WI.

22. Quality Control and Quality Assurance

Quality assurance (QA) is the establishment of systems that ensure that a quality product will be obtained. It involves aspects such as design of plant and processes that will enable a product to be produced "right first time," e.g., design of pipelines that don't harbor microorganisms and gentle filling of tanks to avoid excess air pick-up. It focuses on the application of techniques such as auditing to highlight process and plant weaknesses.

Quality control (QC) involves the measurement of parameters and the subsequent response to those measurements.

Take for example microbiological control:

The QC approach is to take samples from a product in package, plate out the sample on agar, and count the colonies growing after several days. If there is growth, then the beer is held or withdrawn from trade (if already shipped) or the problem ignored.

The QA approach involves personnel from all parts of a plant being involved in the design of a quality operation, including:

- vessel and pipeline configuration for ready cleanability
- effective operation of CIP systems
- rapid methods for assessment of hygiene status of tanks before filling
- good maintenance and operation of pasteurizers.

The QA department provides methodology to assess performance (e.g., caustic strengths in CIP systems, rapid microtechnology) and provides audits and performance measures.

Engineers would be involved in design and fabrication. Brewers operate the system to best quality practice. Wherever possible, measurements would be made by operators.

The QA department performs specialist analysis and acts as the independent "conscience" of the operation. Increasingly breweries around the world incorporate quality systems such as ISO 9000, which formalize the operation as it pertains to quality output.

The establishment of specifications for raw materials, process, and product demands the availability of methodology to make the necessary measurements. Wherever possible (at least on a large scale), sensors should be installed to make measurements automatically linked to feedback-control systems.

Several fundamental measurements are needed for control of malting and brewing, notably

- temperature
- mass
- volume
- pressure

Table 24. Analyses that should be made and responded to for the brewing process to be kept under control

Parameter	Methodology
Absence of taints in liquor supply	Taste it daily
Specific gravity of wort collected in brew house and when fermenter is filled	Hydrometer or "vibrating" U-tube instrumentation
Dissolved oxygen in wort pre-yeast dosing	Oxygen sensor
Amount of yeast "pitched"	Sensors based on light scatter or permittivity
Vicinal diketones in freshly fermented beer	Spectrophotometry or gas chromatography
Alcohol content of beer for declaring tax and controlling dilution of high-gravity brews	Various, including distillation, gas chromatography, infrared or near-infrared spectroscopy
Gases (CO_2, O_2, N_2) in bright beer	Specific gas sensors
Clarity of bright beer	Haze meter
Color of bright beer	Spectrophotometer, Tintometer, tristimulus colorimeter
Bitterness of bright beer	Spectrophotometer, high-performance liquid chromatography
Parameters during packaging (alcohol, gases, color, contents, integrity of seams between can and lid)	Contents by weighing; physical stripdown and visual examination for seam leaks
Strength of CIP detergent	Analyze NaOH concentration
Flavor acceptability	Taste contents of all package types

Table 24 lists the minimum analytical regime that is necessary in a brewery.

Brewers cooperate to establish the methodology used to measure their products, raw materials, and processes. Otherwise there will be little hope of achieving agreement between supplier and user (e.g., maltster-brewer; brewer-retailer) or in circumstances such as franchise brewing.

The principal standardized methods are those of the ASBC and the EBC. Relevant methods are debated in committee and then written up in a standardized format that is clearly understandable so that the methods can be faithfully pursued in whichever laboratory uses them. The method and samples for measurement are circulated to a wide range of laboratories, which individually produce a set of data.

This is collated and analyzed statistically by ASBC, which is able to assign values for repeatability (r) and reproducibility (R). "Little r" (as it is called) is a measure of how consistent the results are when a method is applied by the same analyst in a single location. "Big R" is an index of how good the agreement is when a method is applied to the same sample but in different laboratories with different analysts.

Only if these values are acceptably low will any confidence be placed in a method for its ability to give reliable and reproducible values that can be used not only for process control but also as a basis for transactions. If values for r and R are good, then the method will be added to the recommended list of methods.

The methods can be classified into chemical analysis, microbiological analysis, and organoleptic analysis.

Chemical Analysis

Alcohol

Perhaps the most critical measure made on beer is its content of alcohol, defining as it does the "strength" of a beer. In many countries (although the United States is not one of them), duty is levied on the basis of alcohol content.

Most commonly, alcohol will be measured by gas chromatography, but other methods may include refractometry and specific alcohol sensors.

Accurate measurement of alcohol is also necessary to control the strength of beer produced in high-gravity brewing. In many breweries, this control is carried out in-line. A sensor prior to the dilution point measures the alcohol content continuously and regulates the rate of flow of beer and water at the subsequent mixing point. The alcohol-measuring sensor may be based on one of several principles, one of the most common being near-infrared (NIR) spectroscopy.

Contents

Allied to the declaration of alcohol, the brewer must also satisfy weights and measures legislation by quantifying the weight or volume of beer in the package. This is generally established on a container-by-container basis by weighing the vessel, be it a keg, can, or bottle. Application of statistical distribution analysis indicates whether the inevitable spread of weights across a population of containers is within acceptable limits.

Carbon Dioxide

The level of CO_2 will be measured in the bright beer tank, most frequently using an instrument that measures CO_2 on the basis of pressure and temperature.

Original Extract and Residual Extract

Allied to the measurement of alcohol, *original extract* is an indicator of the strength of a product. If the alcohol content of a beer is known, it is possible to calculate the quantity of fermentable sugar that must have been present in the wort prior to fermentation. This can be added to the *real extract* (which is sometimes called the *residual extract* and which comprises nonfermented material, primarily dextrins) to obtain a value for the original extract. The real extract is determined as specific gravity using a hydrometer or, more commonly, a gravity meter. The latter operates on the basis of vibrating a U-tube filled with the beer. The frequency of oscillation relates to how much material is dissolved in the sample. The real extract tells the brewer whether the balance of fermentable to nonfermentable carbohydrate in the wort was correct and whether the fermentability of the wort was too high or too low.

pH

pH is measured using a pH electrode.

Color

Color is routinely measured from the absorbency of light at a wavelength of 430 nm. For many products, there is a reasonable correlation between this value and color, but it is by no means absolute. The perception of color by the human eye depends on the assessment of absorption at all wavelengths in the visible spectrum. The modern standard for color measurement employed in many industries is based on "tristimulus and chromaticity," which basically describes color in terms of its relative lightness and darkness and its hue (Fig. 112). The color of beer can also be assessed by comparison with that of each of several discs in a device called a Lovibond Tintometer.

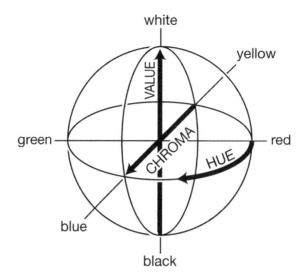

Fig. 112. The CIE Lab approach to color description. A sample is positioned in three-dimensional space and its color quantified in terms of its lightness/darkness, intensity of absorbance in blue-yellow and green-red dimensions, and hue. In this way, samples can be matched against reference standards to get a precise assessment of their perceived color. By contrast, measurements at a single wavelength (usually 430 nm) offer only a measure of intensity of light absorbance.

Clarity

Haze is measured in beer by the assessment of light scatter by particles. Traditionally this has been by shining light through the beer and measuring the amount of light scattered at an angle of 90°. The more light scattered, the greater the haze. For most beers there is good agreement between the amount of light scattered in this way and the perceived clarity of the product. Sometimes a beer may contain extremely small particles that are not readily visible to the human eye but which scatter light strongly at 90°. The beer looks bright, but the haze meter tells a different story. This phenomenon is called "invisible haze" or "pseudo haze." It doesn't present a quality problem in the trade, but it is highly problematic for the brewer, who is forced to make a qualitative judgment as to whether a beer rejected instrumentally is satisfactory for release to trade after all. Nowadays there are haze meters that read light scatter at 13° rather than at right angles, and these don't pick up invisible haze. Unfortunately, they also miss some of the bigger particles.

Dissolved Oxygen

One of the causes of haze formation and flavor deterioration could be a high level of oxygen in the package. Reliable measurement of oxygen is essential, and this is generally carried out using an electrode based on principles of electrochemistry, voltametry, or polarography (Fig. 113). It must be performed before any pasteurization, for the heating will "cook in" the oxygen.

Prediction of Stability

Oxygen is only one factor that will influence the physical breakdown of a beer. The most common building blocks of a beer haze are proteins and polyphenols. As yet, nobody has proved which proteins in beer are particularly prone to throw hazes, and, until

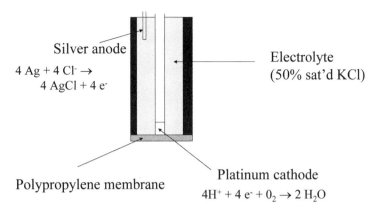

Fig. 113. Oxygen electrode.

this is rectified, the only way to test the level of haze-susceptible protein is to "titrate" them out. In some quality control laboratories, samples of beer will be dosed with aliquots of either ammonium sulfate or tannic acid. The more of these agents needed to precipitate out protein and throw a haze, the less haze-forming protein is present. Many brewers measure the other components of haze, the polyphenols. These can be quantified by measuring the extent of color formation when beer is reacted with ferric (iron) ions in alkaline solution. Although this measures total polyphenols, and they are not all harmful (for instance some are likely to be antioxidants), it is a very useful means for checking whether a polyphenol adsorbent such as PVPP has done its job. Most frequently, beer stability is forecast through the use of breakdown tests. Beer may be subjected, for instance, to alternate hot and cold cycles to try to simulate storage in a more-rapid timeframe.

Bitterness

Most brewers rely on a method that extracts the bitter iso-α-acids from beer with iso-octane, measuring the amount of ultraviolet light that this solution absorbs at 275 nm. The greater the absorbency, the greater the bitterness. Reduced iso-α-acid preparations absorb less light pro rata to their perceived bitterness. Therefore, for a beer bittered with these materials, the factor used is different from when "conventional" hops are used. If the equation is bitterness = $A_{275} \times 50$ for normal hops, it might be $A_{275} \times 70$ for beers incorporating the reduced preparations. High-performance liquid chromatography (HPLC) enables quantitation of the individual isomers of the iso-α-acids.

Diacetyl

A colorimetric method is available to measure diacetyl, but more frequently it is assessed by gas chromatography. It is important to measure not only free diacetyl but also its immediate precursor, acetolactate. Before the gas chromatography, therefore, the beer is warmed to break down any precursor to diacetyl.

Other Flavor Compounds

Some brewers will measure other flavor contributors as well, but, for all brewers, it is through smelling and tasting the beer that they will make their key assessment of its acceptability and judge whether it can be released to trade. Trials have been undertaken with so-called "artificial noses," sensors that are claimed to be able to mimic the human olfactory system. They are far from ready for the job. Among the volatiles that the brewing quality control lab may be required to measure, by gas chromatography, are acetaldehyde, dimethyl sulfide, and a range of esters and fusel oils. It is most likely that this will be on a survey basis, perhaps monthly, rather than brew by brew.

Foam Stability and Cling

By measuring the carbon dioxide content, the brewer has an index of whether a beer has sufficient capability to generate foam. This will not tell whether the resultant foam will be stable, for which another type of analysis is necessary. This is a difficult task, and there is much debate over the best way to measure foam stability. In the United States, there is reasonable acceptance of the Sigma value test as the recommended method. The Institute of Brewing and Distilling does not seem to feel that any method is worthy of recommendation. The two most frequently used procedures worldwide are those of Rudin and NIBEM.

Metals and Other Ions

Several inorganic ions are measured in the brewery, mostly on a survey basis. The cations are measured by atomic absorption spectroscopy. Liquid chromatography is used to detect the levels of a range of anions.

Microbiological Analysis

Traditionally, microbiological analysis in breweries consisted of taking samples throughout the process and inoculating them on agar-solidified growth media of various types designed to grow specific categories of bacteria or "wild yeasts" (i.e., any yeast other than the one used to brew the beer in question). When the plates were incubated for 3–7 days, any organisms on them would grow to produce colonies: the more colonies, the greater the contamination. The problem is that by the time the results were made available and discussed with the brewer, that particular batch of wort or beer would have long since moved on to the next stage. Any remedial procedures would help subsequent brews only.

Far-sighted brewers now use a QA approach to plant hygiene, allied to the use of rapid microbiological methodology. Much more attention is given to designing the plant for easy cleanability, checking the efficiency of cleaning (CIP) systems (e.g., caustic checks), and confirming that the pasteurizer is working by testing temperature and applying various checks to determine whether heat-sensitive components are being destroyed.

Various rapid microbiological techniques have been advocated. The most publicized and most widely used is based on "ATP bioluminescence." The method depends on the firefly, an insect that emits light from its tail as a mating signal. This reaction depends on an enzyme called luciferase, which converts the chemical energy store found in all organisms (ATP) to light energy. The enzyme can be extracted and this reaction carried out in a test tube. The more ATP present, the more light is produced, and it can be measured using a luminometer.

The rapid test used by brewers requires that a swab be scraped across the surface to be tested. The end of the swab is then broken off into a tube that contains an extractant, together with the luciferase; and, after a period that can be as short as a few minutes, the amount of light emitted is measured. The dirtier the surface, the more ATP will have got onto the swab and, in turn, the more light will have been measured. And so, in real time, an indication of the state of hygiene of the plant can be obtained. The method has been extended to measuring very low levels of microorganisms in beer, enabling the brewer to release beer to trade with confidence just a few hours (or, at most, days) after it was packaged.

Sensory (Organoleptic) Analysis

Although the drinking of beer is a complex sensorial experience, bringing into play diverse visual stimuli and environmental factors, ultimately it is the smell and taste of a beer that will decide whether or not it will prove acceptable to the consumer. For this reason, much time and effort is devoted within the brewery to the tasting of beer at all stages in its production.

In well-run breweries, brewers will taste not only the beer freshly packaged but also raw materials and process streams, in order that any flavor defect can be picked up at the earliest possible stage and before the defective material has passed to the next stage in the process. Beer should be tasted at least at the cold-conditioning stage, at the post-filtration stage, and after packaging. The water that is used to brew beer and to dilute high-gravity beer should be assessed. It is, of course, important that the tasters be sensitive to all the flavors expected and unexpected in each product. In practice, several tasters should make decisions on a sample.

Difference Tests

As the name suggests, these tests are intended to tell whether a difference can be perceived between two beers. For instance, the brewer may be interested in checking whether one batch of beer differs from the previous batch of the same beer, whether a process change has had an effect on the product, or whether batches of the same brand of beer brewed in two different breweries are similar, etc.

It is essential that the tasters not be distracted in this task. The environment has to be quiet, and they must not be influenced by the appearance of the product; so the beer is served in dark glasses, in a room fitted with artificial red light, and with no opportunity for them to make contact with other assessors. It is important that the sensitivity of the tasters not be influenced by their having recently enjoyed a cigarette or a coffee or partaken of any strongly flavored food. It is best to have the tasting session prior to lunch.

The classic difference procedure is the three-glass test: seven assessors, at a minimum, are presented with three glasses. Two of the glasses contain one beer, the third the other beer. The order of presentation is randomized. All the taster has to do is indicate which beer he/she thinks is different. Statistical analysis will reveal whether a significant number of tasters is able to discern a difference between the beers and, therefore whether, according to the law of averages, the public will or will not perceive two beers as tasting different.

Descriptive Tests

The three-glass test can be carried out essentially by anyone. However, if a brewer wants to have specific descriptive information about a beer, he/she must use trained tasters, people who are painstakingly taught to recognize a wide diversity of flavors, to articulate about them, and to be able to "profile" a beer. Tasters are trained over time, using a series of defined standards. The terminology is often represented in the form of a "flavor wheel" (see Fig. 5 in Chapter 4). A group of individuals will taste a selection of beers, scoring the individual attributes, perhaps on a scale from 0 (character not detectable) through 10 (character intense). Obviously, it takes real ability to be able to separate out the various terms and recognize them individually, without one parameter influencing another. Once the scoring is complete, the individuals will discuss what they have found and agree on how the flavor of a beer should be summarized.

This type of test is widely used to support new product development, brand improvement, and, of course, to characterize the beers from a competitor. Once again, there are variants of it, such as the trueness-to-type test. This procedure is well suited to assessing whether a beer brewed in one brewery is or is not similar to the reference (standard) beer brewed in another location. For each of various terms found in the flavor profile form, each assessor is asked to mark whether the sample has the same degree of that character (score = 0), slightly more (+1), substantially more (+2), slightly less (–1), or substantially less (–2). Obviously the more flavor notes that are judged to have a score of 0, the more similar are the two beers.

Further Reading

Bamforth, C. W. (2002) Standards of Brewing: A Practical Approach to Consistency and Excellence. Brewers Publications, Boulder, CO.

23. Environmental Impacts and Outputs

Beer is not the only output from a brewery. For instance, the unutilized portion of the grist (spent grains) needs to be disposed of quickly, as it is prone to spoilage. More yeast is produced during fermentation than can be used to re-pitch successive fermentations. There is trub to be dealt with. Large quantities of water with varying amounts of dissolved solids are produced, for instance, in cleaning and cooling. Forty-five percent of total water use in a brewery is for cleaning operations.

Table 25 lists some of the main impacts on the environment from a brewery.

There are occupational health issues in the brewery: noise, caustic and acid, ammonia, heavy loads, dust, wet floors, glass, forklifts, and ventilation issues surrounding CO_2.

Resource Consumption and Emissions

Table 26 shows input and output data for a high- and a low-consuming brewery.

Conscientious brewers address issues such as heat, water, and electricity consumption, plotting their performance brewery by brewery, setting targets, organizing working parties, etc. Heat consumption might be improved by attending to issues such as insulation, steam condensate return systems, boiler efficiency, and heat-recovery systems, etc. Water consumption may be minimized by attending to leaking valves, running taps and hoses, bottle-washer efficiency, enhanced pipeline design, improved CIP efficiency, etc., etc. Electricity waste occurs with inefficient equipment, equipment continuing to operate when unnecessary, leaky compressed air systems, poor insulation and building sealing, too high a condensing temperature, or too low an evaporating temperature in the cooling plant, etc.

Table 25. Sources of environmental impact in a brewery

Operations	Brew House	Fermentation/ Downstream	Packaging	Ancillary
High discharge of organic matter	x	x	x	
High energy use (heating or cooling)	x	x	x	x
High water use	x	x	x	x
Dust	x	x		
Heavy loads	x		x	
Caustic/acid from CIP	x	x	x	
Solid waste	x	x	x	x
Noise			x	x
Air pollution				x
Handling chemicals				x
Hazardous waste				x
Ammonia				x

Spent grains are the residual solids after mashing. The aim of mashing and wort separation is to generate as much extract as possible in wort, provided there is not the excess leaching of undesirables (e.g., excess tannin). If extract is left with spent grains, this is a waste. In a well-run brewery, the difference between the extract measured in the lab on a small-scale batch of malt and that obtained in the brewery should be <1%. If it is higher, then extract is being lost with the grains.

Spent grains are typically about 20% solids, meaning that they are extremely wet and need to be moved rapidly out of the brewery, because they will rapidly become infected. It is obvious that the grains must be discharged from a lauter tun or mash filter quickly so as not to hold up production. In the vast majority of locations, grains are taken away "as is" in trucks and used for cattle food or after ensiling (where lactic acid bacteria are allowed to take the pH down for preservation). Most beer (and therefore spent grain) is produced in the summer, whereas in many countries, the grains are needed as a cattle feed in the winter when grass is not accessible. Shipping in trucks results in seepage of water from the grains (which will contain some extract) out onto the road.

The approximate composition of dried grains is shown in Table 27. The fiber represents the lignocellulose and hemicellulose in the husk. N-free extract will include soluble polysaccharides and any extract not efficiently removed in wort separation.

Spent grains are seldom dried commercially, as the additional cost is prohibitive when balanced against the sales value. They may be partially dried. A range of potential uses for grains has been proposed, including use in breakfast cereals, cookies, snack foods, growing mushrooms, etc.

Table 26. Inputs and outputs

	High Consumption	Low Consumption
Inputs		
Malt/adjunct (kg)	18	15
Energy (MJ)	350	150
Electricity (kWh)	20	8–12
Water (hl)	20	5
Outputs		
Beer (hl)	**1**	**1**
Waste water (hl)	18.5	3.5
Biological oxygen demand (BOD) (kg)	1.2	0.8
Spent grains (kg)	17	14
Yeast (kg)	3	3

Table 27. Composition (%) of dried grains

Component	Percent
Water	8
Fiber	18
Protein	21
Nitrogen-free extract	40
Fat and oil	9
Ash	4

Weak wort produced at the end of the wort separation operation is a significant contributor to biological oxygen demand (BOD) if it is not recovered and used for mashing-in the next brew.

Trub is the precipitate generated in the boiling stage, and apart from protein and solid hop material, it will contain entrained wort. In an effective whirlpool, the amount of trub is 0.2–0.4% of the wort volume, and the trub comprises 15–20% solids. Some brewers discharge the trub directly to drain, in which case it comprises a significant BOD factor (110,000 mg/liter). Others mix the trub with spent grains, which must be performed with good mixing; otherwise, unpalatable (to cows) slugs of bitter material result.

Yeast not required for the re-pitching of subsequent fermentations can be put to various uses. It some countries, it is sold to companies who autolyze it and convert it into extracts that are typically spread on toast. Other brewers sell it as pig food after it has been killed with propionic acid. It may be sold to distillers, for whom the specific products of fermentation are less important than they are for brewers. Typically 2–4 kg of yeast (10–15% dry matter) is produced per hectoliter of beer. This constitutes a very high BOD loading of 120,000–140,000 mg/liter.

There are a number of sources of *residual beer* in a brewery, which may account for up to 1–5% of total production. Generally, this is recovered and blended, where possible, provided the quality of the main beer stream is not jeopardized. Otherwise, it goes to drain, with a BOD of 80,000 mg/liter.

Some sources of recovered beer are:
- beer remaining after emptying of tanks
- pre-runs and post-runs from filters. At the beginning of filtration, a filter is full of water, which is pushed out with beer. At the end of filtration, the beer is pushed out with water. The result is a mixture of beer and water.
- beer pushed out of pipes with water
- beer rejected in the packaging operation (e.g., wrong fill heights, quality defects, label misalignment)
- returned beer
- exploding bottles (e.g., bottle defects)
- beer with solid additions used for stabilizing beer

Some other materials. Other materials include kieselguhr (100–300 g/hl of a diminishing resource), leading to suggestions of regeneration and the advent of improved filtration control to lower use of kieselguhr (the tendency is always to overdose). Cross-flow filtration has also been investigated.

The *caustic* consumption of a brewery is about 0.5–1.0 kg of 30% NaOH per hectoliter. High consumption may be due to poor recovery during CIP and to problems with the bottle washer. If waste water is not neutralized, then caustic increases its pH.

Fermentation produces approximately 3–4 kg of carbon dioxide per hectoliter of wort. Some brewers recover this, scrub it, and use it as a motor gas in their breweries. Others vent it to atmosphere. CO_2 is also generated in the boilers (16 kg/hl of beer) when burning fossil fuel, and it is not usually recovered.

Table 28 lists the solid waste from a brewery (other than grains and yeast).

Finally, breweries produce odor—notably from the brew house. Few people find this offensive, although there have been questions about the level of volatile organic com-

Table 28. Solid waste (kg/10^3 hl beer) from a brewery

Material	Amount
Broken glass	850
Treatment sludge	100–800
Labels, paper	290
Filter aid	250
Cardboard, cartons	40
Plastics	50
Metals	20

pounds (VOCs) emitted from a brewery. However, in reality, the problem of VOCs in brewing is probably not very great. In the United Kingdom, for instance, the whole of the UK food industry emits 74,000 tons per annum, which is less than 4% of industry total. Of the food and drink output of VOCs, less than 5% is from malting and brewing; and of that in malting and brewing, 95% is emitted in malting.

Wastewater Treatment

Some brewers send their wastewater (and sundry materials dissolved and suspended in it) directly to local treatment works. Others are obliged to treat water themselves, using either aerobic or anaerobic digesters. In the latter case, they may generate a worthwhile supply of methane for fuel use.

The Mogden formula emerged from the United Kingdom to quantify effluent treatment costs:

$$C = R + V + (V_b \text{ or } V_m) + (O_t/O_s)B + (S_t/S_s)S$$

where
- C = Total cost per m³ for treatment and disposal of effluent
- R = Reception and conveyance charge per m³
- V = Volumetric and primary treatment charge per m³ for those effluents discharged to a sewage treatment works where there is no biological treatment
- V_b = As for V, but additional cost for biological treatment
- V_m = As for V, but where discharge is to sea outfall
- O_t = Strength of settled effluent as determined by an oxidation parameter (usually COD) (mg/liter)
- B = Biological oxidation charge per m³ of settled sewage
- S_t = Suspended solids content of the trade effluent in mg/liter
- S = Treatment and disposal charge for primary sludge per m³ of sewage
- O_s = Mean strength of settled sewage at a sewage treatment works in mg/liter
- S_s = Mean suspended solids content of sewage at a sewage treatment works in mg/liter

Further Reading

Anonymous. (1996) Environmental Management in the Brewing Industry. Technical Report 33. United Nations Environment Programme, Paris.

Appendix 1

Chemistry and Biochemistry for Brewers

Proteins

Proteins perform a number of critical functions in nature. These include
- structural roles (e.g., the collagen found in many tissues, including the swim bladders of fish that end up as isinglass finings)
- the carrying of other molecules around (e.g., the transport proteins that have a role in selectively moving substances across membranes, such as the membrane surrounding brewing yeast)
- imparting mobility (e.g., in the whiplike flagella found on certain spoilage microorganisms)
- protection, as antibodies that are produced by higher organisms in response to the presence of "foreign" materials (such antibodies are increasingly widely used in selective tests for certain materials of interest to the brewer)
- catalysis (the enzymes, including those from malt and yeast, are critical in the brewing and fermentation processes)

Amino Acids

Proteins are polymers (see Some Necessary Chemistry) of amino acids. Amino acids are so named because they have a (basic) amino group ($-NH_2$) and an (acid) carboxyl group ($-COOH$). There are 20 of them commonly found in proteins, and they share a general formula:

$$\begin{array}{c} NH_2 \\ | \\ H-C-COOH \\ | \\ R \end{array}$$

Each amino acid has a different "R" group. These are shown in Figures 114 and 115. The simplest, in glycine, comprises a solitary H atom. Others, such as that in tryptophan, are quite complex. The properties of these R groups impact greatly on the properties of the individual amino acids but also on the properties of the proteins in which they are found. The oddity is proline, which in the strictest terms is not an amino acid, but it obviously has close similarities to the amino acids. It is an imino acid.

-H	Glycine		-CH$_2$OH	Serine
-CH$_3$	Alanine		-CH(OH)CH$_3$	Threonine
-CH(CH$_3$)$_2$	Valine		-CH$_2$SH	Cysteine
-CH$_2$CH(CH$_3$)$_2$	Leucine		-CH$_2$CH$_2$SCH$_3$	Methionine
-CH(CH$_3$)CH$_2$CH$_3$	Isoleucine		-CH$_2$COOH	Aspartic acid
			-CH$_2$CH$_2$COOH	Glutamic acid

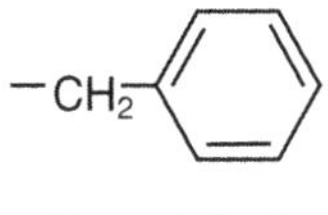

-CH$_2$—[phenyl]   Phenylalanine

-CH$_2$—[phenyl]—OH   Tyrosine

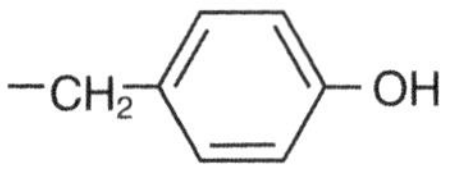

Fig. 114. The side-chains in amino acids (part one).

-CH$_2$CONH$_2$	Asparagine
-CH$_2$CH$_2$CONH$_2$	Glutamine
-CH$_2$CH$_2$CH$_2$CH$_2$NH$_2$	Lysine
-CH$_2$CH$_2$CH$_2$NHCNH$_2$NH	Arginine

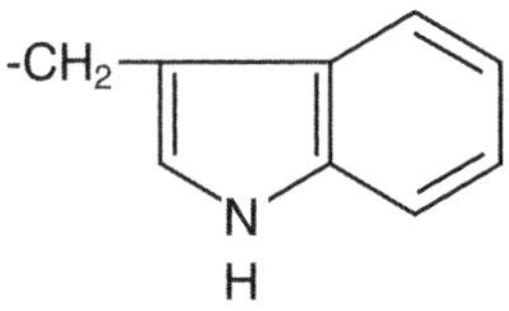

-CH$_2$—[indole]   Tryptophan

Histidine (imidazole side-chain: -CH$_2$-C=CH with NH and NH$^+$ connected via CH)

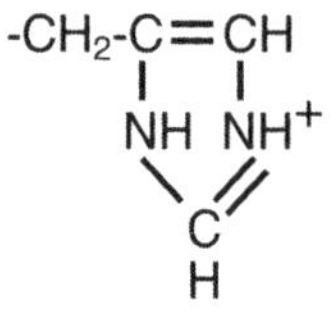

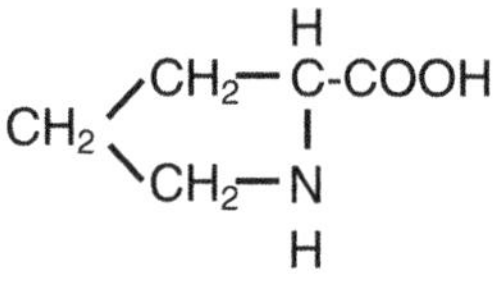

Proline (cyclic structure with CH$_2$—C(H)-COOH ring via CH$_2$—N—H)

Fig. 115. The side-chains in amino acids (part two).

Both the amino group and the carboxyl group of amino acids can become charged (see Some Necessary Chemistry). Carboxyl is an acidic group; i.e., it releases hydrogen ions (H$^+$). Amino is a "basic" group because it accepts hydrogen ions.

If the pH is very low (high concentration of H$^+$), then the carboxyl group tends to be in the uncharged form (–COOH), whereas the amino group tends to be in the positive ionic form (–NH$_3^+$). The converse applies at high pH. For intermediate pH levels, the majority of the molecules will have both a positive and a negative charge—they are *dipolar*.

$$\underset{\text{low pH}}{\begin{array}{c} NH_3^+ \\ | \\ H-C-COOH \\ | \\ R \end{array}} \leftrightarrow \underset{\text{neutral}}{\begin{array}{c} NH_3^+ \\ | \\ H-C-COO^- \\ | \\ R \end{array}} \leftrightarrow \underset{\text{high pH}}{\begin{array}{c} NH_2 \\ | \\ H-C-COO^- \\ | \\ R \end{array}}$$

This ability of amino acids to alternately "soak up" and release H⁺ means that (within limits) they are able to regulate the pH of solutions that contain them—i.e., they are *buffers*. As such, they play an important role in regulating the pH of wort and beer.

Peptide Bonds

If two amino acids come together, then a molecule of water can be split out between them and a bond formed that joins them together. This is called a "peptide" bond (bold in the diagram):

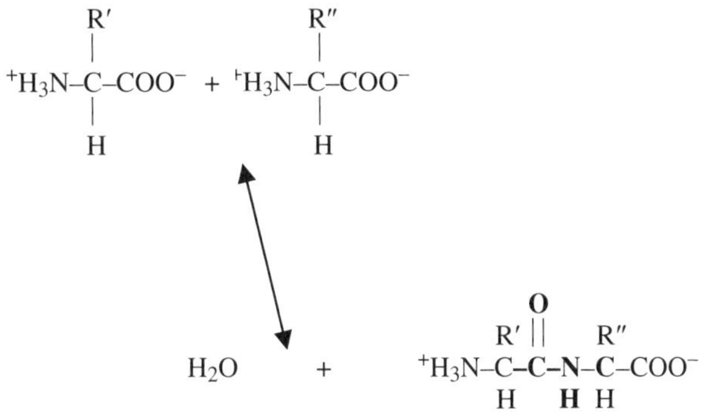

$$^+H_3N-\underset{H}{\overset{R'}{C}}-COO^- + {}^+H_3N-\underset{H}{\overset{R''}{C}}-COO^-$$

$$\updownarrow$$

$$H_2O \quad + \quad {}^+H_3N-\underset{H}{\overset{R'}{C}}-\overset{O}{\overset{\|}{C}}-\underset{H}{N}-\underset{H}{\overset{R''}{C}}-COO^-$$

The product is a *dipeptide*. This reaction is reversible: when water is added, the dipeptide splits into its corresponding amino acids.

There are a very large number of possibilities for dipeptides. For any two amino acids, there are two dipeptides. Take glycine and alanine, for instance. The first dipeptide has the peptide bond formed between the –COO⁻ of glycine and the –NH₃⁺ group of alanine. The second has the peptide bond formed between the –COO⁻ of alanine and the –NH₃⁺ of glycine. They are quite distinct, having their own properties. Convention has it that peptide structures are written with the free (unbound) amino group to the left and the free carboxyl group (the one not tied up in a peptide bond) to the right. Thus, the first of our dipeptides would be glycyl-alanine. The second would be alanyl-glycine.

There is huge scope for more and more amino acids to be joined on to the chain. Broadly speaking, if there are between two and approximately 10 amino acids in the chain, then we have a *peptide* or *oligopeptide*. If there are more, then we have a *polypeptide*.

In these polymers, only one of the amino acids has a free amino group (apart from the amino groups in the side-chains of certain amino acids), and this is called the *N-terminus*. Similarly, only one amino acid has a free carboxyl group (apart from the carboxyl groups in the side-chains of glutamate and aspartate), and this is called the *C-terminus*.

Most polypeptides found in nature contain between 50 and 2,000 amino acids. The average molecular weight of an amino acid is approximately 110, so the molecular weight of polypeptides may be as much as 220,000 (220K). (It seems to have become the norm to insert the word "daltons" after molecular weights, in honor of the Manchester-based professor who did the pioneering work on atomic structure. This is strictly incorrect, as molecular weights are unitless. They are ratios of size on a scale relative to an atom of carbon, with an atomic weight of 12.)

The Importance of Amino Acid Sequence

The sequence of amino acids in a polypeptide chain is called its *primary structure*. It is this that determines the properties of the polypeptide, for the simple reason that the *folding* of the polypeptide chain is determined by the opportunity for amino acids in different regions of the chain to interact. This folding is referred to as *secondary* and *tertiary* structure.

Various forces dictate these interactions. There may be strong attractions between positively and negatively charged groups in the side-chains of the amino acids—opposites attract; i.e., positive attracts negative. Equally, like charges repel, so that positive repels positive and negative repels negative. And thus the impact of this on the shape of a protein molecule depends on how many of the amino acids are present that have these charges on their side-chains. It also depends on what the pH is, as this influences the proportion of the groups that are present in a charged form or an uncharged form (see earlier).

If the conditions are right, then it may be that amino acids quite far apart in the molecule come close together and are "cemented" by these strong ionic interactions (Fig. 116). In this way, the polypeptide chain adopts a certain shape.

Other interactions can occur. Some of the most important are called *hydrogen bonds*. Suffice to say, in the chemical grouping containing one oxygen and one hydrogen (i.e., –OH, the so-called *hydroxyl* group) the oxygen atom is slightly negatively charged and the hydrogen atom is slightly positively charged. Therefore, if two separate –OH groups come close together, there can be an interaction between their respective positive and negative regions (Fig. 117). Individually, these hydrogen bonds are weaker than the full-

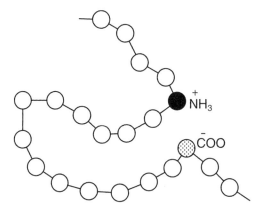

Fig. 116. Ionic bonds.

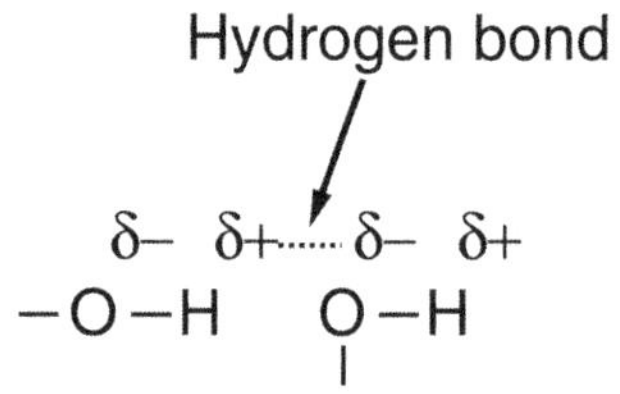

Hydrogen bond

$\delta-\ \ \delta+\cdots\cdots\delta-\ \ \delta+$
$-O-H\ \ \ O-H$

Fig. 117. Hydrogen bond.

fledged positive-negative bonds referred to above, but cumulatively they can be very significant. Similar bonds can occur with N-H groups. In fact, hydrogen bonds involving –OH and –NH groups are responsible for the coiling of some regions of proteins rather in the way that the cord linking a telephone receiver to the main unit twists into a helix. Alternatively, these bonds can link adjacent regions of proteins into sheet forms.

This hydrogen bonding is most marked in water, where of course there are many –OH groups, and whole network structures can be formed between adjacent molecules. It's a very cozy situation—and a party very difficult to break up by intruder molecules and groups. Examples of the latter types of groups are the amino acids with hydrocarbon side-chains (i.e., those amino acids with side-chains entirely composed of C and H). They cannot make hydrogen bonds and so they are "excluded from the club." In fact, they tend to be banished to the inside of protein molecules where they associate together. They are referred to as *hydrophobic* amino acids because of this "water-hating" tendency. It is the *hydrophilic* groups that predominate on the outer surface of the proteins, because they can "pal on" with the water molecules—for example, the –OH group of serine is perfectly able to hydrogen bond with water.

Another type of bond that determines the shape of protein molecules is the so-called *disulfide bridge*. Amino acids containing –SH side-chains may come together, and the hydrogen atom of each is removed as the two sulfur atoms link. The bond formed is even more resistant to breakage than an ionic bond, and in some proteins these bridges are very important for determining the overall shape.

Figure 118 gives a hypothetical picture of how these various types of interactions determine the shape of a protein molecule. You need to try to imagine these things happening in three dimensions. We are dealing with essentially spherical structures, i.e., baseballs rather than frisbees.

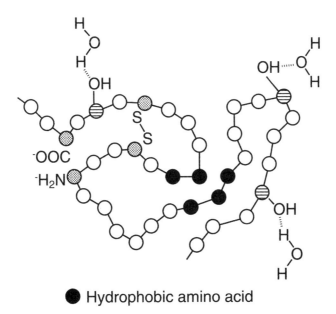

Fig. 118. Interactions responsible for the three-dimensional structure of polypeptide chains.

Why Does Protein Structure Matter to the Brewer?

Protein structure matters for various reasons.

Enzymes

Enzymes are proteins, each of which is capable of effecting its own specific reaction. Amylases exclusively break down starch; β-glucanases break down β-glucans; proteinases break down proteins, and so on.

Enzymes are biological catalysts—molecules present in all living organisms that are responsible for speeding up the reactions occurring in those systems (see Some Necessary Chemistry). The molecules they act on are called substrates; the molecules formed are products.

The more enzyme available, the faster the reaction. We can liken them to turnstiles at a football ground: the more there are, the more quickly customers can get through. The relationship between the speed of an enzyme-catalyzed reaction and substrate concentration is not so simple, and the relevant graph usually has a shape like that shown in Figure 119. At a certain substrate concentration, the system becomes "saturated": increasing the substrate concentration no longer causes the reaction to go any faster. The

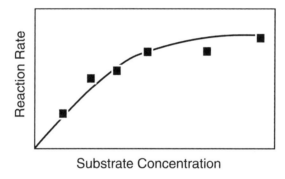

Fig. 119. Relationship between substrate concentration and reaction rate in an enzyme-catalyzed reaction.

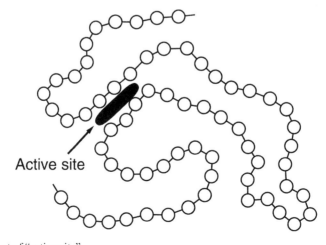

Fig. 120. The concept of "active site".

explanation is that the enzyme binds to the substrate molecule to form an "enzyme-substrate complex." This then breaks down to reform the enzyme (ready to grab another substrate) and release the product. This binding occurs at the so-called "active site." There is a unique shape in each enzyme molecule that recognizes the substrate. The shape of the protein molecules determines this niche by the way in which it folds on itself through the interactions described earlier. It's rather like a space module docking with the mother ship. A U.S. module needs a U.S. docking port, not a Russian one (by and large!).

The active site might comprise amino acids from quite distinct parts of the enzyme molecule (Fig. 120). As enzymes are relatively flimsy, anything that tends to drive these amino acids apart disrupts the active site, prevents substrate binding, and destroys enzyme activity. Such factors include heat, which causes the protein to jiggle about until it loosens up, and changes in pH, which alter the proportions of positive and negative groups in the proteins and therefore the opportunity for the charge-charge interactions that hold the shape of the molecule. These are the reasons that extremes of temperature and pH are to be avoided. Enzymes differ in their tolerance of temperature and pH.

In addition to their effects on the integrity and "survival" of the protein molecule, temperature and pH also directly affect the rate of the reaction that the enzyme catalyzes. All chemical reactions are accelerated by heat, with the rule of thumb being that a 10 degree C rise in temperature speeds up a reaction by two- to threefold. This applies to reactions catalyzed by enzymes, with the important caveat that increasing heat will also tend to disrupt the enzyme structure, deform the active site, and therefore prevent the enzyme from doing its job. Thus, the net rate of reaction observed is a balance dependent on how resistant the enzyme is to heat. Enzymes in mashes such as α-amylase and peroxidase are very resistant to heat, whereas others like β-glucanase and β-amylase are much more heat sensitive.

The pH, too, has impacts other than on the stability of the enzyme. We need to remember that the active site comprises several amino acids drawn from different parts of the overall polypeptide (and perhaps even from several polypeptides—some enzymes contain two or four separate polypeptide chains and are said to have quaternary structure). Some of these amino acids, such as glutamic acid or lysine, have side-chains that can become charged (see earlier). It may be, for instance, that the enzyme uses the negative charge of a glutamic acid residue to bind the substrate. If that were the case, too low a pH, which would make the glutamic acid uncharged, would lead to a cessation of enzymatic activity.

Furthermore, some substrates can exist in uncharged or charged forms and may need to be in one or other form in order to become attracted to the enzyme. In fact, most enzymes operate effectively only within a narrow pH range. Most enzymes of relevance in mashing happen to work best in the pH range of mashes (between 5 and 6).

It is not only changes in pH that can interfere with enzyme activity and stability. Enzymes are also susceptible to inactivation by other agents (*inactivators*). One such substance is the copper ion (Cu^{2+}). This snares –SH groups and inactivates irreversibly those proteins that depend on –SH groups for their function.

Other molecules that can block enzyme activity do so reversibly (i.e., if you take them away enzyme activity is restored), and they are known as *inhibitors*. Basically, there are four types of inhibition that we need to concern ourselves with. Remarkably, although each almost certainly occurs in brewery systems, they haven't been studied in great detail on an enzyme-by-enzyme basis.

The first, *competitive inhibition*, results when a molecule is so similar to the substrate molecule that it can "recognize" the active site and bind to it but is sufficiently different that it just sits there and cannot be tackled by the enzyme. It gets in the way of the substrate. However, if the substrate concentration is high enough, then it will squeeze out the inhibitor, and inhibition is overcome.

In *noncompetitive inhibition*, the inhibitor binds to a site on the enzyme that is not the active site and, in so doing, distorts the shape of the molecule so that the active site is no longer available to substrate. Removing the inhibitor allows the enzyme to slide back into shape, but in this case increasing the substrate concentration cannot squeeze out the inhibitor.

At very high substrate concentrations, we sometimes encounter *substrate inhibition*. Once again we can turn to a football stadium for an analogy: if the crowd is dense outside and not forming an orderly line but rather jockeying to get through a single turnstile, then nobody will enter comfortably. So it is with substrate inhibition: separate substrate molecules interfere with one another's ability to bind to the active site.

The fourth significant type is *product inhibition*. In this case, it is departing product molecules that interfere with the substrate's ability to bind—rather as if a single turnstile was being used to let people leave the stadium as well as bring them in.

Interactions of Proteins with Other Molecules

So we see that it is not only substrate molecules that can interact with proteins. Let's consider two examples relevant to the brewer.

The first interaction is between the bitter compounds (iso-α-acids) and polypeptides. The interactions involved here are of at least two types (Fig. 121). In the first place, the hydrocarbon hydrophobic parts of the iso-α-acids and those of polypeptides link. In turn, the negative charges on adjacent iso-α-acids are bridged by the two positive charges on a

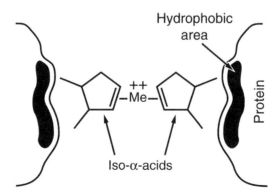

Fig. 121. Iso-α-acid–metal ion–protein interaction in foam. The iso-α-acids are extremely stylized, depicting only two of their three hydrophobic arms, how these hydrophobic arms interact with hydrophobic regions in polypeptides, and how the negative charges in adjacent iso-α-acids are neutralized by bridging through a divalent metal cation (such as magnesium, manganese, or zinc). In this way, large complexes between separate proteins and iso-α-acids are formed and strength is imparted to bubble walls.

metal ion, such as magnesium. The network formed stabilizes proteins in bubble walls (i.e., in foam). The proteins that tend to give more stable foams are relatively hydrophobic. From understanding the rudiments of protein structure (as we have explored here), we can explain why: hydrophobic groups want to get together and get away from water. (In reality, it's because the hydrogen-bonding water fraternity excludes them.) They associate together (and with other hydrophobic molecules, such as the bitter acids) and migrate to surfaces away from the body of the water (or beer), i.e., to the bubble walls.

The second interaction is between proteins and polyphenols. The latter tanning molecules have a shape that fits comfortably with proline (Fig. 122), and haze-forming polypeptides are rich in proline. Furthermore, it is hypothesized that each polyphenol can bind to more than a single polypeptide. These interactions build up and build up until large networks are produced that are so big that they are no longer soluble. The result is haze. In turn, silica hydrogel is effective in removing haze proteins and polyvinylpolypyrollidone in removing polyphenols because they have surfaces that are similar to polyphenols and proline-rich proteins, respectively.

Solubility

As we have seen, the delicate structure of proteins that presents a hydrophilic exterior with the hydrophobic groups on the inside ensures solubility but is easily disrupted. One of the main factors in brewing that will do this is heat. Thermal energy "blows" the structure wide open, the water-hating interior is exposed, and these hydrophobic regions search one another out with the formation of insoluble clumps. This is what happens in wort boiling. Similarly, proteins are sensitive to cold—the molecules rearrange and become insoluble (cf., chilling before filtering).

Another agent that can come to bear is oxygen. Adjacent –SH groups on proteins can be oxidized to form bridges:

$$-SH + -SH + O_2 \rightarrow -S-S- + H_2O_2$$

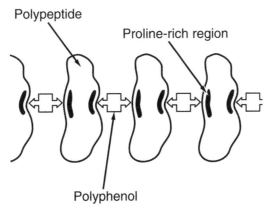

Fig. 122. Polyphenol-protein interactions in haze.

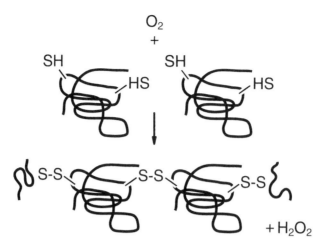

Fig. 123. Oxidation of sulfhydryl groups in proteins.

In this way, separate protein molecules can join together and reach a size that makes them insoluble (Fig. 123). In a mash, this type of reaction contributes to the formation of *teig* and impedes wort flow in lautering; the more oxygen present, the greater the opportunity for this to happen.

Carbohydrates

Carbohydrates have many roles in life. Quantitatively, they are the major organic materials in nature (*organic* refers to compounds containing carbon atoms, other than the simpler ones such as carbon dioxide).

Food Stores

First, carbohydrates comprise the major food reserves in living systems. Two examples are the starch found in plants (for example, in the endosperm of barley) and the glycogen found in animals—and brewer's yeast. Starch and glycogen are polymers of glucose: huge numbers of separate molecules joined together to form long chains. If you completely break down starch or glycogen, you end up with glucose.

The reason an organism keeps its food reserves in this polymeric form is to avoid problems with osmotic pressure. If a concentrated solution of glucose (many glucose molecules per milliliter of water) is separated from a weak glucose solution by a membrane that allows passage of small molecules, then water will pass across the membrane in an attempt to equalize the strength of the solutions on both sides of the membrane. If that membrane were surrounding a cell and that cell were loaded with glucose as a store, then water from the outside of the cell would come flooding in and the cell would swell and eventually burst once the membrane had been overstretched. However, if all those glucoses are joined together to make a molecule of starch or one of glycogen, then the concentration of molecules is much lower. It's the concentration of individual molecules that determines osmotic pressure, not the size of those molecules. So by maintaining glucose in a polymeric form, a cell prevents osmotic stress.

Source of Energy

Why store carbohydrate anyway? The answer is that it can be "burned" as a fuel to generate energy. The pathways that do this are discussed later. Suffice to say here, the storage materials (starch, glycogen) are broken down by enzymes to form simpler sugars, which then enter into quite complex pathways, each involving a separate enzyme-catalyzed step, that progressively, in a highly controlled and efficient manner, leads to the capturing of energy.

Forming a Part of Important Metabolic Molecules

The energy is captured in the form of a high-energy molecule, called adenosine triphosphate (ATP). ATP can be likened to a crossbow. When you pull back the string and secure it, it is akin to the pathway by which sugar is broken down and which leads to the accumulation of ATP—i.e., you are putting energy into the crossbow. Then when you release the catch, the energy is released, hopefully in a targeted fashion. For ATP this means releasing its energy again in metabolic processes that require energy, such as making cell components, moving about, and simply staying alive. The ATP is broken down.

ATP has quite a complex structure. Remarkably, at its heart is a sugar, called ribose (see later). This is an example of a sugar being part of important metabolic molecules. Ribose is also found in RNA (ribonucleic acid), while a ribose lacking one oxygen, ergo deoxyribose, is an important feature of the structure of DNA (deoxyribonucleic acid), the genetic material of the vast majority of living cells. (RNA helps DNA translate its code into action, as we shall see later.)

Structural Roles

Carbohydrates have diverse important structural roles in nature. Crab shells are carbohydrates. Closer to home, perhaps, are the β-glucans and pentosans in the starchy endosperm cell walls of barley, the cellulose in the husk, and the glucans and mannans in the cell wall of yeast (more on this later).

Carbohydrates also form a part of the structure of some proteins. Some people are adamant that the foaming polypeptides are so-called glycoproteins, the glyco- prefix signifying that sugars are present.

The Chemistry of Carbohydrates

Some Basics

The word "carbohydrate" derives from the fact that many of these compounds have the general formula $C_n(H_2O)_n$; i.e., they are "hydrates of carbon" (hydrate as in hydration—inclusion of water). So glucose, one of the simpler sugars, has the simple formula $C_6H_{12}O_6$, where n = 6.

The simplest carbohydrate of all is glyceraldehyde. This comprises a carbon atom to which are attached four different groups: a simple hydrogen atom (–H), a hydroxyl group (–OH), an aldehyde group (–CHO), and an alcohol group (–CH$_2$OH). These "poke out" at equal angles from the central carbon.

To get a feel for this you may wish to avail yourself of a small potato (the carbon atom) and four cocktail sticks (Fig. 124). Push the cocktail sticks into the potato such that

200 / Appendix 1

each is equidistant from all the others. The cocktail sticks are the bonds that bind the carbon atom to each of the groups that are linked to it. Now get a cocktail onion and stick it on to one of the sticks. You will instantly realize that this is your hydrogen atom. Similarly, we can put onto the other sticks a gherkin (aldehyde group), a cherry tomato (hydroxyl), and baby carrot (alcohol). (Please don't get hung up on the size issues here—for instance, the fact that the potato carbon is much bigger than the carrot, which, being our alcohol, not only has a carbon but also three hydrogens and an oxygen. The fruit and vegetables are being used in an illustrative manner only!)

You can, of course, move this "fruity vegetables" version of a glyceraldehyde molecule about in space, and the onion, gherkin, tomato, and carrot always stay in the same places relative to one another, no matter what group is uppermost.

However, let's switch the sticks holding the onion and the tomato (Fig. 125). What we have now is the mirror image of our first molecule. And it's different! No matter what you do (short of whipping out the sticks or switching the fruit and vegetables), you can't turn one around such that it looks exactly like the other.

What we have are *isomers*: molecules that differ in the orientation of the groups. There are, then, two forms of glyceraldehyde: they are known as D-glyceraldehyde and L-glyceraldehyde (Fig. 126). (Remember them by thinking of the L as meaning "left," such that the –OH points to the left when the aldehyde group is pointing upward.)

Fig. 124. A simple way to understand stereochemistry: D-glyceraldehyde.

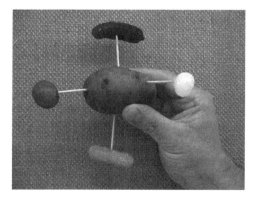

Fig. 125. L-Glyceraldehyde.

In fact, the carbohydrates that primarily concern brewers are all in the D- configuration, so we'll concentrate on that class.

Now suppose we put another carbon atom between the central carbon of glyceraldehyde and the aldehyde group, and on this new carbon we have a hydrogen (onion) and a hydroxyl (tomato). The reader will realize (I hope—if not, get some more cocktail sticks and food) that the new addition can be introduced in two ways: in one the onions (and tomatoes) are on the same side of the molecule, whereas in the other they are on opposite sides. Thus, we have two new molecules, which are in fact called D-erthyrose and D-threose (Fig. 127).

In this way, we can keep adding more H–C–OH or HO–C–H segments (Fig. 128). And each time, because we have two ways in which to make the introduction, we double the number of possible compounds.

Glyceraldehyde has three carbons—it is a *triose*. The suffix *-ose* indicates carbohydrate: most carbohydrates have this suffix. Those sugars with four carbons are known as tetraoses ("tetra" meaning four). Then we have the pentoses, hexoses, etc.

Polymers of sugar units containing five carbons are known as pentosans. Polymers of glucose are known as glucosans, or more commonly, glucans.

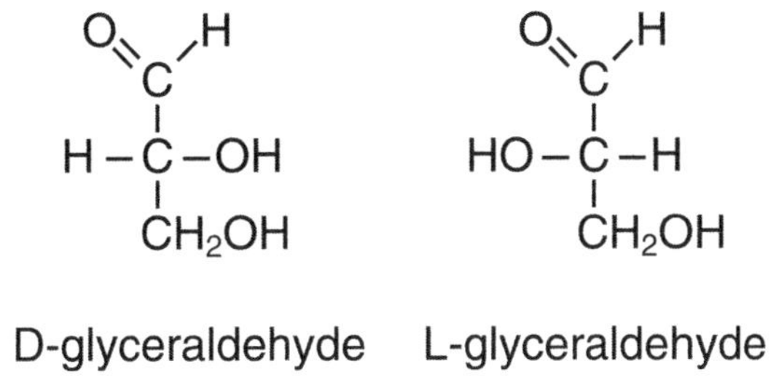

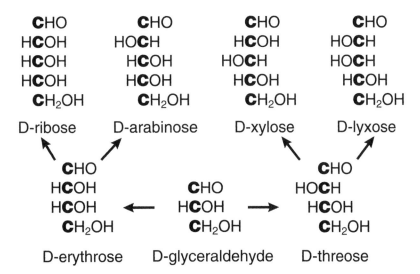

Fig. 126. Structures of glyceraldehydes.

Fig. 127. Every time you add a carbon atom, you double the options.

202 / Appendix 1

Figures 127 and 128 show that all of the molecules are very similar. Small differences, though, can make all the difference in this world. Thus, mannose and glucose differ only in the way in which two of the groups are attached to one of the carbon atoms (in fact, carbon atom number 2—the convention for numbering the carbons is shown in the glucose molecule). This is a tiny difference, but one that makes for a different set of properties for glucose and mannose, for example, their sweetness, the enzymes that act on them, etc.

Building Up the Complexity

In fact, a sugar such as glucose has a more complex structure than this. For the most part, it is found in a ring form, rather than a linear style. Think of one end of the molecule grabbing the other, as a snake might grip its jaws on its tail end.

In fact, there are two ring forms that are interchangeable through a linear form (Fig. 129). These two forms are called α and β. In the α-form, the –OH group at C-1 points downward, and for β it points upward.

$$
\begin{array}{cc}
^1\text{CHO} & \text{CHO} \\
\text{H}^2\text{COH} & \text{HOCH} \\
\text{HO}^3\text{CH} & \text{HOCH} \\
\text{H}^4\text{COH} & \text{HCOH} \\
\text{H}^5\text{COH} & \text{HCOH} \\
^6\text{CH}_2\text{OH} & \text{CH}_2\text{OH} \\
\text{D-glucose} & \text{D-mannose}
\end{array}
$$

Fig. 128. The two D-hexoses derived from arabinose.

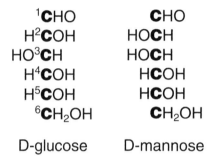

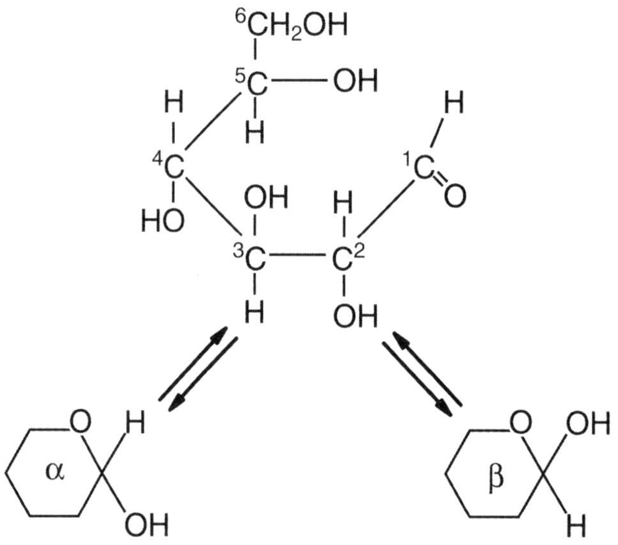

Fig. 129. Glucose in linear and two ring forms.

Carbon atom number 1 is also called a *reducing group*, because when the molecule is in the linear form it has a free carbonyl group (–C=O), which is readily oxidized (and when something is oxidized something else is reduced). In other words, a molecule such as glucose is able to reduce another molecule because of this group. It is through this group that a sugar interacts with an amino acid (see later) to form colored materials through the so-called Maillard reaction that occurs in kilning and wort boiling. If these reducing groups are eliminated (for example by polymerization—see next section), then color formation will not occur.

Joining Sugars Together

Low molecular weight carbohydrates tend to be freely soluble in water and, of course, sweet. They are called sugars. A molecule of glucose (or xylose or mannose, etc.) is called a *monosaccharide*.

Individual sugar molecules can join. If their hydroxyl groups get together, a molecule of water can be split out and they share the remaining oxygen atom. A sugar comprising two such units is called a *disaccharide*. Add another unit and we get a *trisaccharide*— and so on. Carbohydrates containing between two and, say, 10 units are called *oligosaccharides*. Anything bigger is a *polysaccharide*.

By now you will, of course, have realized that two glucoses (for example) could join in various ways. One of the most common ways is for the water to be split out from between carbon 1 and carbon 4. In this way, we can get a 1-4 linkage (an example of a *glycosidic bond)*. However, there are two possibilities, depending on whether the glucose providing the –OH from its C-1 atom is in the α or β configuration. Thus, we can have an α 1-4 or a β 1-4 linkage (Fig. 130). The two disaccharides formed are totally different. The first is maltose (the main sugar found in wort); the second is cellobiose.

When hundreds and hundreds of glucose units are linked together through α 1-4 bonds, we have a molecule called amylose, which is one of the two fractions found in starch (Fig. 131). The other is called amylopectin, which differs from amylose in that it has branches that result from some of the glucoses being linked α 1-6.

When loads of glucose units are linked together through β 1-4 bonds, the molecule is cellulose. Just one small change in configuration (β links between the 1 and 4 carbons rather than the α ones) makes a huge difference in properties. Cellulose is the major component of most plant cell walls, in tree trunks, for instance, ending up in paper. It is tough and difficult to chew. The β configuration between two glucoses allows the production of very straight chains. Not only that, there is an opportunity for hydrogen

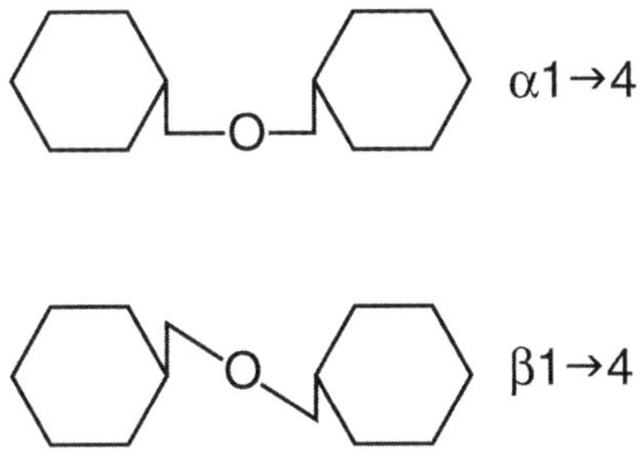

Fig. 130. Two glucoses can link by either α or β bonds between carbon atoms 1 and 4.

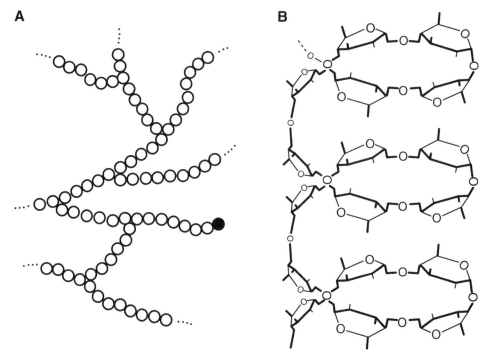

Fig. 131. A, Amylopectin, where the various chains join in an α 1-6 bond. ● indicates the sole non-reducing end. B, Amylose (linkages are α 1-4).

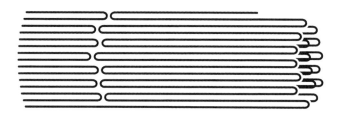

Fig. 132. Cellulose fibrils.

bonding to increase yet further the interaction between adjacent glucoses. Hydrogen bonding between bundles of adjacent chains leads to fibril formation (Fig. 132).

Contrast this with starch—familiar in everyone's kitchen. (You wouldn't bake your cakes from paper.) The α links make for a helical structure (Fig. 131) akin to a loose spring. You will realize that we have a somewhat open, accessible structure.

The cell walls of barley endosperm are about three-quarters β-glucan. It wouldn't be sensible for this to be cellulose, because the barley embryo needs to break down its food reserve readily, and it would take an eternity to get through cellulose-rich walls. Thus, a third of the inter-glucose links in this case are β 1-3 (Fig. 133). Such links (which for the most part occur interspersed by two or three β 1-4 bonds) tend to disrupt the ability of the polymer to fold in on itself and make those hydrogen bonds between side-by-side glucoses. However, there are still plenty of –OH groups poking out, and they can interact with their look-alikes on adjacent chains. In this way are built up aggregates that increase

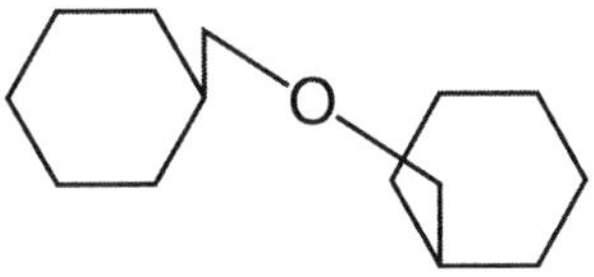

Fig. 133. Laminaribiose (linkage is β 1-3).

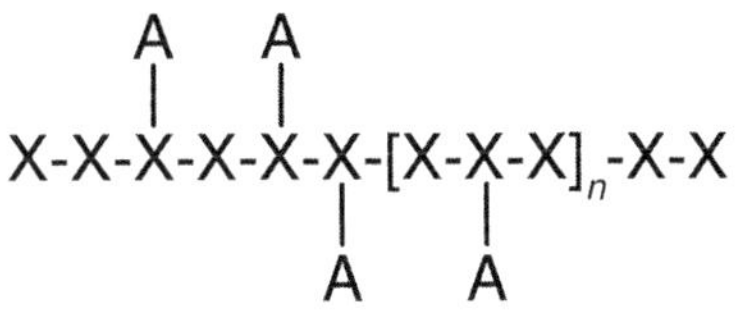

$$\begin{array}{c} A \quad A \\ | \quad | \\ X\text{-}X\text{-}X\text{-}X\text{-}X\text{-}X\text{-}[X\text{-}X\text{-}X]_n\text{-}X\text{-}X \\ | \quad | \\ A \quad A \end{array}$$

Fig. 134. Arabinoxylan. X = xylose; A = arabinose; $[\,]_n$ = multiple residues.

the viscosity of solutions and reduce the solubility of the molecules (ergo, the brewer encounters problems with β-glucans).

The other polysaccharide in the cell walls of the starchy endosperm of barley is the pentosan arabinoxylan (Fig. 134). In this case we have long chains of xyloses joined through β 1-4 bonds. Arabinoses are joined to some of the xyloses by α 1-2 and α 1-3 links.

The glucans and the pentosans in the barley walls are robust enough to provide structural integrity while being sufficiently amenable to enzymic digestion. An organism such as barley needs to develop at least one different enzyme for every type of glycosidic linkage that needs to be broken down. In this way, the organism can control the breakdown of its food reserve: if one enzyme broke down everything, we'd have chaos. The hydrolysis involves the addition of water (i.e., the restoration of the water that is split out when two monosaccharides *condense* together to form a disaccharide, and so on).

Some of these enzymes are said to be *endo* acting, in that they attack a polysaccharide molecule in the middle to produce smaller oligosaccharides. Others are *exo* enzymes, which attack from the end of a molecule, chopping off one or two sugar units. These tend to travel in only one direction and not from both ends of the molecule. In fact, an enzyme such as β-amylase attacks from the so-called *non-reducing end*. Take a look at the amylose molecule (Fig. 131). You will recall that all the links between the glucoses are α 1-4. The involvement of carbon atom number 1 in making these linkages means that the free reducing group of glucose is eliminated. Only one glucose unit retains its reducing power—the one to the far right that is not involved in forming a glycosidic bond. This is the *reducing end*, which is at the opposite pole to the non-reducing end.

Other Groups Sticking to Carbohydrates

A range of other, noncarbohydrate groups can link to carbohydrates. One such is phosphate. As we shall see when we encounter metabolic pathways later, when a sugar such as glucose is attached to phosphate it becomes "activated."

Polymeric carbohydrates may also have phosphate groups attached to them. An example is phosphomannan, a polysaccharide found in the outer surface of the yeast cell wall. As the name suggests, this is a polymer of mannose, to which is attached phosphate groups. The latter, being able to display a negative charge, give the wall its charge.

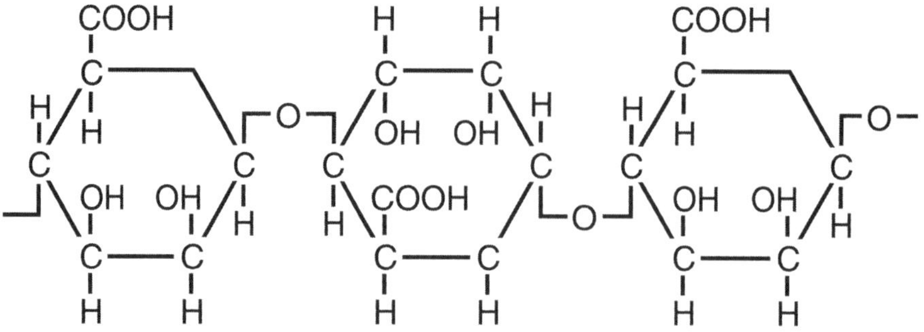

Fig. 135. Alginic acid.

In fact, the yeast wall also contains a β-glucan. It's not cellulose, and neither is it β 1-3, β 1-4-glucan as in barley. Rather it contains β 1-3 and β 1-6 linkages.

Another polysaccharide material not unknown to brewers is alginate. It has been used as a clarifying or fining agent because of its negative charge and ability to grab positively charged proteins, the attendant "lump" being too big to stay in solution with the result that it precipitates out. Alginate has the formula shown in Figure 135. It is possible to convert this into a different form, known as propylene glycol alginate (PGA), by the addition of propanol. The propanol (an alcohol) reacts with the acid (–COOH) groups. The product of the reaction of an acid and an alcohol (with the splitting out of water) is called an *ester*. (Simple smaller esters like *iso*-amyl acetate are important flavor constituents of beer.) PGA, of course, has protective powers in the realm of foam stability.

An ester linkage to a polysaccharide that is native to barley is that between ferulic acid and the arabinoxylan in the endosperm cell wall. There is growing evidence that this may play a role in limiting the degradability of the walls.

Lipids

Lipid is a term used to describe those chemical compounds in living organisms (such as barley, hops, yeast, and you and me) that are insoluble in water. They comprise a diverse range of molecules, many of which are classed as fats or oils. Closely related to them are other materials that make a sizeable contribution to the lives of most of us, substances like detergents and petrol. Unsurprisingly (because of their insolubility), the lipids are primarily found in structural components of cells, principally the membranes that surround the cells and hold the soluble liquid components within.

Structure of Lipids

Many lipids are based on combinations of fatty acids. These compounds have a basic structure comprising chains of carbon atoms joined end to end, with hydrogen atom "wings" and, at one end, an acid group, the carboxyl group (–COOH) (Fig. 136). The simplest such molecule is actually acetic acid. It has just two carbon atoms, one holding three hydrogens and the other being the carbon in the carboxyl group. Acetic acid is strictly not a lipid because it is very water soluble (being the key ingredient of vinegar). It is only when successively more carbons and hydrogens are added that the insolubility becomes manifest, for reasons that we will discuss momentarily.

Those fatty acids with up to 14 or so carbon atoms are often referred to as "short chain fatty acids." They often have distinctive aromas, such as "cheesy."

In living organisms, the most quantitatively important fatty acids have either 16 or 18 carbon atoms. (Much longer fatty acids are found in some organisms but in progressively smaller quantities).

Of especial interest are the fatty acids containing 18 carbon atoms, of which there are several. The first, stearic acid, has each of its carbon atoms linking through a single bond to its neighboring carbon atoms (Fig. 137). Hydrogen atoms take up the other places on the first 17 carbons, the 18th carbon being part of the carboxyl group. (In terms of

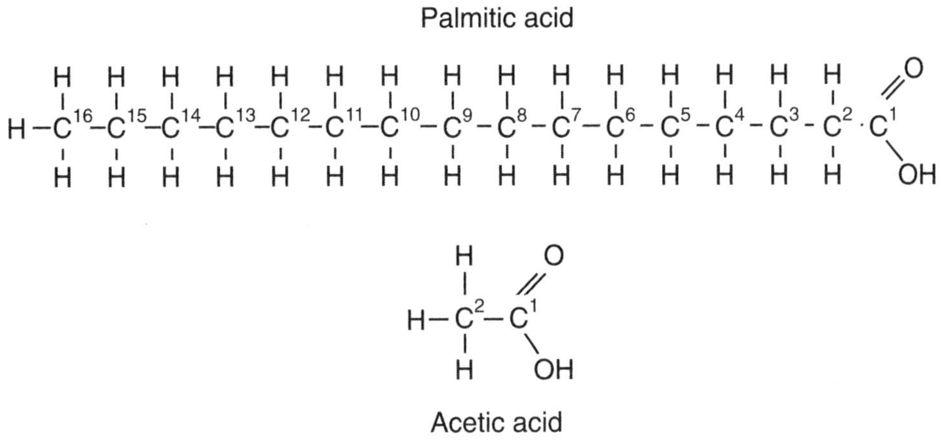

Palmitic acid

$$H-\underset{H}{\overset{H}{C^{16}}}-\underset{H}{\overset{H}{C^{15}}}-\underset{H}{\overset{H}{C^{14}}}-\underset{H}{\overset{H}{C^{13}}}-\underset{H}{\overset{H}{C^{12}}}-\underset{H}{\overset{H}{C^{11}}}-\underset{H}{\overset{H}{C^{10}}}-\underset{H}{\overset{H}{C^{9}}}-\underset{H}{\overset{H}{C^{8}}}-\underset{H}{\overset{H}{C^{7}}}-\underset{H}{\overset{H}{C^{6}}}-\underset{H}{\overset{H}{C^{5}}}-\underset{H}{\overset{H}{C^{4}}}-\underset{H}{\overset{H}{C^{3}}}-\underset{H}{\overset{H}{C^{2}}}\cdot \underset{OH}{\overset{O}{\underset{\diagdown}{C^{1}\nearrow\!\!\!\!=}}}$$

$$H-\underset{H}{\overset{H}{C^{2}}}-\underset{OH}{\overset{O}{\underset{\diagdown}{C^{1}\nearrow\!\!\!\!=}}}$$

Acetic acid

Fig. 136. Basic structures of fatty acids. Carbon 1 is the so-called "carboxyl carbon" because it forms part of the carboxyl group, which can dissociate to release a hydrogen ion (H⁺), leaving –COO⁻.

Fig. 137. The fatty acids with eighteen carbons. Some shorthand conventions are illustrated. Rather than write out all the carbon atoms, biochemists simply draw lines, at each end of which is a carbon atom. As each carbon atom is capable of binding to four other atoms, it is taken as read that the last carbon atom on the left has three hydrogen atoms attached to it, and most of the other carbons have two hydrogen atoms linked to them. The carboxyl group is, of course, different, with two bonds to one of the oxygen atoms and one to the other, which in turn links to another hydrogen atom (c.f. Fig. 136). The unsaturated fatty acids contain double bonds. Those carbon atoms entering into double-bond formation have only one hydrogen atom linked to them. The shorthand notation is to indicate the total number of carbon atoms in front of the colon and then to signify the number of double bonds after the colon. Thus, 18:2 is universally recognized by biochemists as being linoleic acid.

208 / Appendix 1

nomenclature, it is actually numbered C_1—the carbon that is at the other end and attached to three hydrogens and one carbon atom is C_{18}.)

Oleic acid differs from stearic acid in that carbon atom numbers 9 and 10 are held together by two bonds. It is said to be an unsaturated fatty acid, whereas stearic acid is a saturated fatty acid.

Then we have linoleic acid, in which there are *two* double bonds, between carbons 9 and 10 and carbons 12 and 13. Linoleic acid is an example of a polyunsaturated fatty acid, as is linolenic acid, which has a third double bond, this one between carbons 15 and 16.

For the most part, in living organisms fatty acids don't exist in a free form; rather they are attached to other molecules, primarily glycerol (Fig. 138). Glycerol is an alcohol (in fact, you might even say it's three alcohols in one, having three of the alcohol groupings, –OH). Fatty acids are, as the name indicates, acids. When an acid and an alcohol react

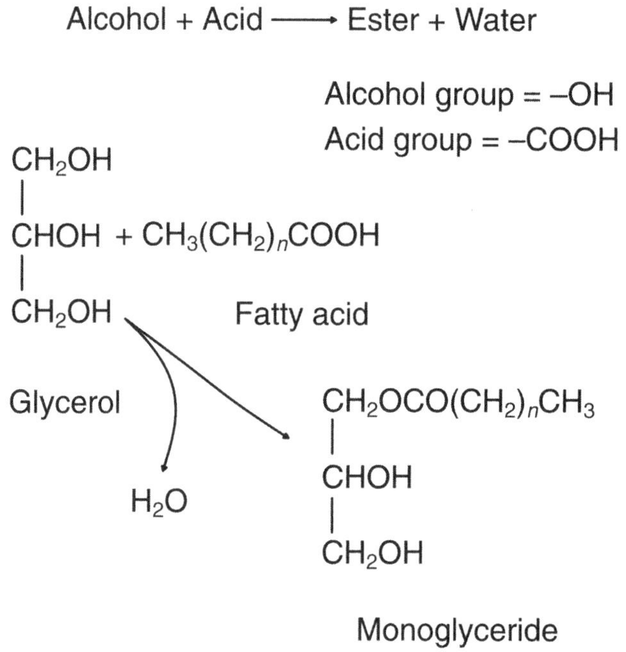

$$\text{Alcohol} + \text{Acid} \longrightarrow \text{Ester} + \text{Water}$$

Alcohol group = –OH
Acid group = –COOH

$$\begin{array}{l} CH_2OH \\ | \\ CHOH \\ | \\ CH_2OH \end{array} + CH_3(CH_2)_nCOOH$$

Glycerol   Fatty acid

$$\begin{array}{l} CH_2OCO(CH_2)_nCH_3 \\ | \\ CHOH \\ | \\ CH_2OH \end{array}$$

H$_2$O

Monoglyceride

Fig. 138. A glyceride is an ester formed between a fatty acid and glycerol.

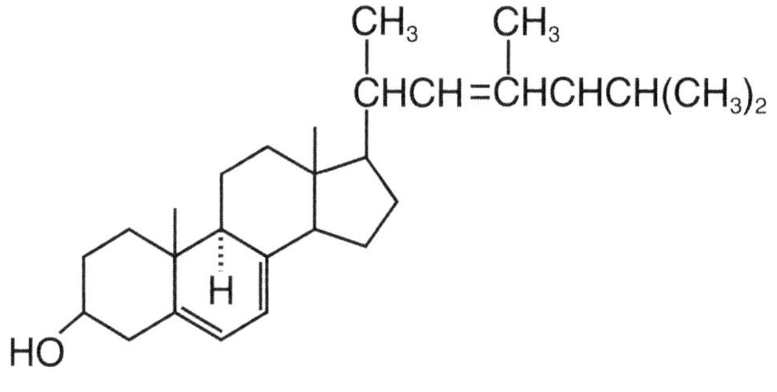

Fig. 139. Ergosterol.

together (by the exclusion of water), an "ester" is produced. Esters of glycerol and fatty acids are known as glycerides. If just one of the –OH groups of glycerol is "esterified" by a fatty acid, we have a monoglyceride. Two fatty acids make a diglyceride, and three, a triglyceride.

There are more complex forms of esterified fatty acids, which contain extra groups, such as phosphate (phospholipids), sugars (glycolipids), and sulfur (sulfolipids). The important properties of lipids that we need to address require us to focus only on the fatty acids and the sterols.

Sterols, alongside the fatty acids, are important components of the membranes of yeast. The most important one is ergosterol (Fig. 139).

Fats and Oils

The shape of such sterols lend themselves nicely to a structural task in membranes. Fatty acids (particularly when in the form of phospholipids) are also major membrane components. In particular, fluidity and flexibility are important.

The structure of a lipid greatly affects its properties. Moving into the kitchen momentarily, we find that butter is a solid (at room temperature) and it is referred to as a fat. The bottled free-flowing stuff is referred to as an oil, because it is a liquid at room temperature. Examples are sunflower oil, corn oil and olive oil.

Clearly, olive oil is much more fluid than butter (unless you heat the butter). Olive oil contains a high proportion of unsaturated fatty acids, which have a lower melting point (the temperature at which they go from being a solid to a liquid) than do the saturated fatty acids found in butter.

The Insolubility of Lipids

To explain the insolubility of lipid we need to consider the forces that hold separate molecules together (see States of Matter). The strongest of these forces are between molecules that have fully fledged charges, positive and negative. Opposite charges attract, so if we have a couple of adjacent molecules, each of which has a full positive and a full negative charge, then the charges on each will attract the opposite charge on the other, and the two molecules will associate (Fig. 140A). An example is common salt (NaCl, which we can also write Na^+Cl^-). To break the units apart demands a lot of energy to overcome the strong intermolecular forces—try heating salt and it will neither melt nor boil.

In some molecules, there are "incomplete" charges. The best example is water. As described earlier, oxygen tends to hog the electrons and pull them away from hydrogen in water, so it has a partial negative charge, whereas hydrogen has a partial positive charge. These partial charges allow adjacent water molecules to associate, through something called "hydrogen bonding" (Fig. 140B). The opportunity exists for vast networks of water molecules to associate to form well-organized structures. At room temperature, water is of course a liquid. The hydrogen bonding needs quite a lot of energy to overcome it, so water only boils at 100°C.

210 / Appendix 1

But what of compounds that don't have any charge whatsoever? An example of such a compound is a fatty acid, such as palmitic acid. (For the purpose of this discussion, let's ignore the –COOH group, which adds complexity to the argument. It is the rest of the fatty acid molecule that largely determines its solubility properties.) Another example is a sterol, such as ergosterol. Such molecules are held together by something called van der Waals forces (Fig. 140C). As we saw earlier, atoms and molecules contain positive and negative charges. In a molecule such as palmitic acid, they balance one another out. However, the individual negatively charged electrons are moving about the whole time, creating infinitesimal localized regions with net positive or negative charges. Association of the tiny opposite charges on adjacent molecules hold them together, but the inter-

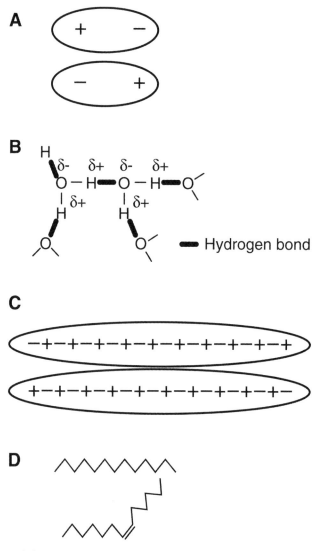

Fig. 140. Interactions and shapes. A, ionic interactions; B, hydrogen bonding; C, Van der Waals forces; D, fatty acids. Saturated fatty acids are straight whereas unsaturated fatty acids are kinked. Imagine stacking corrugated sheets and such sheets after they have been bent.

actions are very weak. Thus some lipids (e.g., sunflower oil) are liquid at room temperature, whereas others (e.g., lard) don't need much input of energy to liquefy them. Whereas the saturated fatty acids have a regular zigzag structure that lends itself to packing together, the double bonds in unsaturated fatty acids put kinks into the molecules, rendering them less easy to pack together (Fig. 140D). Fats, with their higher proportion of saturated fatty acids, therefore are solid, whereas the unsaturated fatty acid-rich oils are liquid.

Another way to overcome the interactions between molecules of the same type is to intersperse other molecules between them, that is, to try to dissolve them. Sodium chloride dissolves in water because the strong interactions between the positively and negatively charged ions can be overcome and replaced by many smaller positive-negative interactions between the sodium and chloride and the water molecules.

This is not possible with something like a fatty acid. A fatty acid does not dissolve in water; it is hydrophobic ("water-hating"). It will dissolve in something that is held together by similar forces, so things like chloroform and petroleum products dissolve fats (as anyone needing to remove a greasy stain from clothing knows). The rule of thumb is "like dissolves like."

Amphipathicity

The carboxyl group in a fatty acid is *hydrophilic* ("water-loving"). In a glyceride, the glycerol portion is the hydrophilic bit. So part of the molecule is hydrophilic and part is hydrophobic. These molecules are said to be *amphipathic* (Fig. 141A).

Now if you add oil (say sunflower oil) to water, the oil will collect atop the water (being less dense). You can't see it, but what is happening is that the hydrophilic parts of the lipids have oriented to dip into the water, where they are quite comfortable, and the hydrophobic portions point away from the water (Fig. 141B).

Within the body of water, lipid may exist in the form of so-called *micelles* (Fig. 141C). Again, the hydrophilic groups gather on the outside, able to interact with water while the hydrophobic chains hide inside.

Structures of the type shown in Figure 141B are exactly analogous to those found in the membranes of living cells, save that the latter tend to comprise a bilayer of lipid, with the hydrophobic chains on the two layers pointing toward each other in the middle, with the respective hydrophilic groups outside.

The action of soaps (and other detergents) in dissolving grease and other hydrophobic "dirt" is analogous to this micelle formation. Soaps are essentially the salts of fatty acids, made by reversing the esterification of glycerides (Fig. 142A). The hydrophobic chains on the soaps can associate with "lumps" of grease and dirt while presenting their charged, hydrophilic ends to the water (Fig. 142B). In this way the unwanted material is solubilized.

Foam

Foams in the kitchen sink are due, of course, to dishwashing liquid (detergent). As the bowl is filled, air is whipped into the system, and the detergent stabilizes the resultant bubbles by forming an interactive surface "skin" that counters the force of surface tension seeking to collapse the bubble (Fig. 104).

212 / Appendix 1

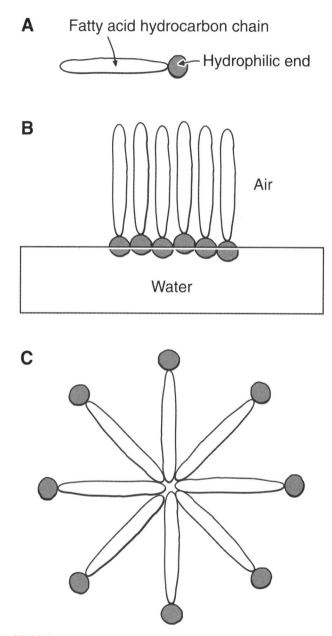

Fig. 141. Behavior of lipids in the presence of an aqueous solvent. A, lipid; B, lipid behavior at a surface; C, lipid behavior (micelle) in the body of the aqueous system.

Of course, lipids and detergents are *bad* news for beer foam. How can this be? The answer lies in the different ways in which beer foams and detergent foams are stabilized. Beer foams are stabilized by interactions between polypeptides that are rich in hydrophobic character (which drives them into the head) and the very hydrophobic bitter acids (Fig. 105). If we have a mixture of protein and detergent or lipid, we have mutual interference. The proteins can't associate together to form a barrier against bubble collapse, and neither can the detergent (Fig. 106). Result—foam collapse.

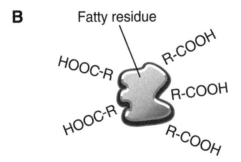

Fig. 142. Soaps and detergents. A, Soaps are formed by alkaline hydrolysis of glycerides; B, soap (or detergent) interacts with the hydrophobic chains associating with the fatty particle and the hydrophilic ends associating with the water.

Lipids and Oxygen

The unsaturated fatty acids are very susceptible to oxidation. Oxygen reacts with the double bonds, starting a cascade reaction leading to the development of materials that can give rancid and cardboard characters to products, including beer. At one time in the manufacture of cooking fats, such materials were industrially hydrogenated to eliminate the double bonds and therefore lengthen the shelf life of the product.

Ironically, oxygen is involved in the metabolic pathways leading to the synthesis of unsaturated fatty acids (and sterols). This is why yeast needs some oxygen to allow it to kick off fermentations efficiently.

Lipids and Energy

Some lipid is found within the amylose component of starch (see earlier); however, most of the lipid in a cereal such as barley is found in the germ. The embryo uses the lipid as a high-concentration energy reserve. As we shall see when we talk about metabolism, living organisms break down fuels using oxygen to release energy. It is exactly analogous to burning coal or wood on the fire. In this combustion process, carbon dioxide and water are produced. Now if you compare the structure of a fatty acid with that of, say, glucose, you will realize immediately that there is already a lot more oxygen pro rata in the glucose molecule than in the fatty acid. In other words, the glucose is already reasonably oxidized. This means that there is more potential energy available from the oxidation of a fatty acid than from a carbohydrate—nutritionists assign a calorific value of 9 kilocalories for each gram of fat but 4 kilocalories per gram of carbohydrate and protein. This is why it makes sense for bodies to store food as fat rather than carbohydrate: you need to carry less weight around. In fact, a 120 pound person would be 150 pounds if there were no such thing as a fat reserve.

Nucleic Acids

There are two types of nucleic acid in living organisms: deoxyribonucleic acid (DNA) and ribonucleic acid (RNA). In all the types of organisms that a brewer is interested in (barley, hops, yeast, and spoilage organisms) the DNA comprises the information store for the cells. RNA constitutes the machinery that transcribes and translates that information into cellular activity.

There are two essential questions surrounding these molecules.
- How is the information passed on from one cell to its daughters when the cell divides, e.g., during growth?
- How is the message in DNA translated into cellular activity; i.e., precisely how does the code get interpreted and acted upon?

For each question to be answered clearly, we need to know the basics of nucleic acid structure.

Structure of Nucleic Acids

Nucleic acids are polymers built of three different types of units: sugars, phosphoric acid, and bases.

The sugars are ribose in RNA (ergo, ribonucleic acid) and deoxyribose in DNA (deoxyribonucleic acid). As we saw earlier, ribose is a sugar with five carbons and five places in its ring structure (i.e., a pentose). Deoxyribose is just ribose missing one of its oxygen atoms (on carbon atom number 2).

Linking carbon atom number 3 on one sugar ring to carbon atom number 5 on the next is the phosphoric acid. In this way, long chains are formed.

Attached to carbon atom number 1 are the bases. DNA and RNA each contain four types of base. In DNA they are called adenine, thymine, cytosine, and guanine. In RNA thymine is replaced by uracil.

RNA molecules comprise individual, single chains, whereas DNA molecules consist of two of these chains intertwined in the form of a double helix (Figs. 143–145). The two chains are linked relatively loosely through an association between the bases on the separate strands.

More specifically, adenine always latches on to a thymine on the opposite chain, whereas cytosine always reaches out to a guanine. These couples fit together snugly and make for a well-ordered molecule.

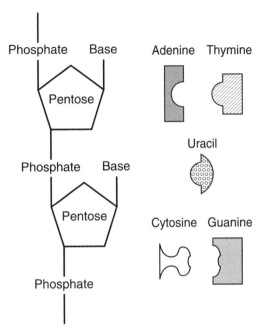

Fig. 143. The basic structural elements of nucleic acids. The configuration of sugars, phosphates, and bases and a schematic representation of the importance of base "shape" to allow fitting of cytosine and guanine or of adenine with either thymine (in DNA) or with uracil (in RNA).

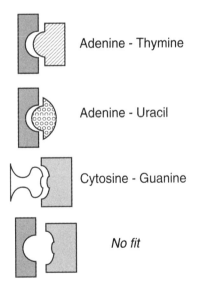

Fig. 144. Allowable fits between bases.

The Importance of Base-Base Recognition

This selectivity of binding of the bases is the very key to the genetic code. It allows us to answer the questions posed earlier.

When a cell divides, the division is preceded by a replication of the entire DNA in the cell. In a process catalyzed by specific enzymes, the double helix is unwound, and to each

216 / Appendix 1

Fig. 145. The pairing of bases on adjacent strands of DNA and an illustration of the double helix. In the type of illustration shown on the left, the vertical lines represent the sugar and phosphate components.

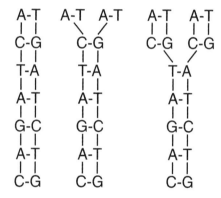

Fig. 146. Replication of DNA. An enzyme starting at one end of the double helix starts to unzip the two strands, and new complementary bases, together with deoxyribose and phosphate, are pieced in.

strand is linked in a new strand, leading eventually to the production of two "daughter" strands, each identical to the "parent" (Fig. 146). You will appreciate that if there is, say, an adenine in position 1 on one strand, then when the new strand starts to be linked in, the machinery knows to attach a thymine to it. If the next base is a guanine then the machinery knows what to do: "ah! cytosine."

In this way, the genetic code is replicated and passed on when a cell divides.

In the same way, the message from DNA is passed on to RNA to initiate the triggering of cellular functions. An enzyme unzips the DNA molecule and, using one of the DNA strands as the template, an enzyme stitches together molecules of RNA. This time, if it sees an adenine it splices a uracil into the chain, and it also pops in ribose instead of deoxyribose.

This process of converting the master code in DNA to the messenger (RNA) is called *transcription*. There are many individual messengers transcribed from a single DNA molecule, each of the messengers carrying the entire code needed for one protein. The

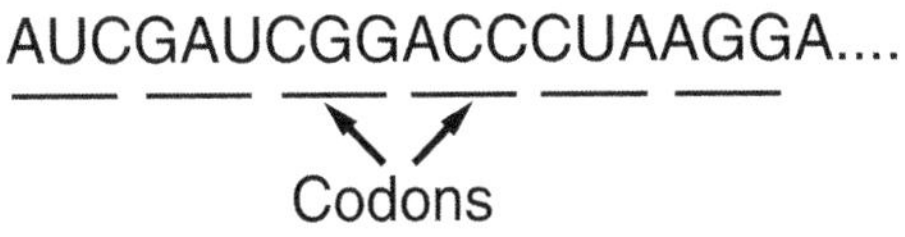

AUCGAUCGGACCCUAAGGA....
Codons

Fig. 147. The basis of the genetic code translated into messenger RNA.

sequence of DNA that is transcribed into an individual messenger RNA molecule is called a gene.

The process whereby the code carried by the RNA is shifted into a protein is called *translation*. The key is the sequence of bases (Fig. 147). Each sequence of three bases indicates a specific amino acid. It is called a *codon*. Thus, for instance, if the reading machinery (another enzyme) sees guanine, guanine, guanine one after the other, then it calls for the amino acid glycine, which is actually delivered by another type of RNA called transfer RNA. (There is one type of transfer RNA for each amino acid) If the next three bases are uracil, cytosine, and guanine, the reading machinery alerts the serine carrier to get itself to the party, and the enzyme splices it onto the glycine. And so on, until the protein molecule is completed.

Everything hinges on the precise sequence of the bases in the DNA. One small change—just in one base—could wreak havoc because the wrong amino acid will be "screwed" into the protein, and this can easily alter the structure of the protein and hence its function

Genetic Modification

DNA is a reasonably sensitive molecule. Its bases can react with various outside influences and become changed. These outside factors include ultraviolet radiation and diverse chemical compounds called mutagens because they mutate the DNA and, as a consequence, influence its behavior. The carcinogens act in this way.

The mutagens might lead to a loss of bases or to a change in bases. For example, nitrous acid converts adenine to something called hypoxanthine, which binds to cytosine and not thymine. When DNA replicates, the new strand will contain cytosine instead of thymine, and this change will carry on to the next generation. The cytosine will now be matched by a guanine on the partnering strand and form a cytosine-guanine pair where there used to be an adenine-thymine pair. When the mutated DNA is translated into RNA, the message passed on is different.

Let's assume that the altered base message in messenger RNA was in a triplet that should have read cytosine-guanine-guanine (Fig. 148A). That codes for arginine. By exchanging the cytosine for uracil, the reading machinery is confused into putting a tryptophan residue into the protein (Fig. 148B). The likelihood is that this will change the resultant protein tremendously, and it probably won't function.

The situation is even worse if the mutation leads to a deletion of one of the base pairs (Fig. 148C). The effect is to shift the reading frame so that all the amino acids coded after the point where the deletion occurred will probably be wrong.

"Natural" mutation is the reason that we're all different and why evolution takes the course it does. The impacts of mutation can be beneficial. If the altered DNA codes for a

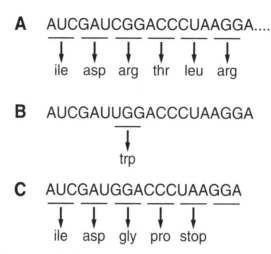

Fig. 148. The impact of mutagens. A, the non-mutated message; B, message obtained in messenger RNA where a base change has occurred (U for C); C, message obtained in messenger RNA where a base (C) has been deleted. Note that UAA is the codon that says "stop reading." Every codon after that point will not be read.

"better" protein—for instance, changing a single amino acid might increase the heat tolerance of an enzyme—then it would be an advantage, e.g., to an organism adapting to growth in high-temperature locations. (This type of thing can be done in the lab—it's called protein engineering.) Of such changes, over a relative eternity, did long necks on giraffes evolve.

Mutation has been used by brewing scientists in the past to eliminate unwanted activities, e.g., the tendency of yeast to develop hydrogen sulfide. And mutation can be used profitably in research studies into the understanding of metabolism.

Cloning is essentially a scissors-and-paste job on DNA (Fig. 149). Take a donor organism and a recipient organism. Extract and chop up (using a certain type of enzyme) the DNA from the former and attach the various pieces to something called a vector. This is frequently either a virus or a bit of circular DNA that can move into and out of cells, called a plasmid. The plasmid-donor DNA is added to the host organism, which has had its wall removed to make it more penetrable. Inside the cell, the bits of DNA (including the gene we are interested in) will either integrate into the host chromosomes or will remain in the vector form, provided there is a positive pressure to keep them there. That might take the form of tagging on to the plasmid some other factor. A popular one is the gene for copper resistance. If the host cells are then grown in the presence of a high level of copper, then only the cells containing the plasmid will survive—and they will also possess the gene we are interested in. Alternatively, the host organism might be one used because it is deficient in something, for example, the ability to use a certain amino acid because it lacks a certain enzyme. The plasmid or vector would then be engineered to possess the gene for the missing enzyme and so the recipient organism becomes able to use that amino acid if it has taken up the plasmid/vector.

The cells that have successfully received the gene of interest (and other genes necessary for reasons just mentioned) are purified. If they are microbes, they can be used (law permitting) straight off. If they are cells derived from plants, there is the issue of regenerating the whole plant.

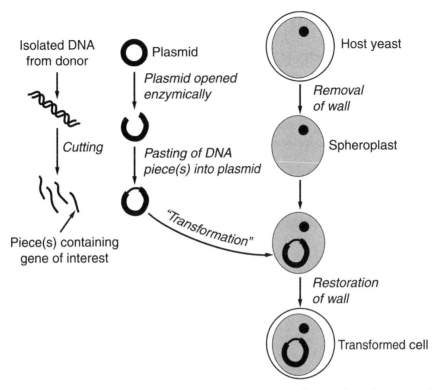

Fig. 149. Recombinant DNA technology. The wall of the yeast cell to be transformed is removed by using selected enzymes (the wall is a barrier to DNA entry). Other enzymes are used to "chop" the donor DNA into pieces. The gene of interest will be on one of these pieces. The various pieces are mixed with an opened ("cut") small carrier ring of DNA called a plasmid under conditions where the ends of the individual DNA pieces and the opened ends of the plasmid will seal together. The plasmid originates from the yeast. This plasmid has certain properties that allow it to be retained by the yeast cell to which it is introduced. The wall-free yeast is mixed with the pool of plasmids. Those yeast cells that have received the DNA of interest (and no other) can be selected. The wall is then regenerated.

It is easier to manipulate the DNA of simple organisms, like bacteria, than DNA from yeast because of ploidy. If you try to isolate a mutant of a polyploid organism you have problems: if you knock out one gene, there is at least one other copy of the same gene to fall back on. And the shortage of mutants makes it a challenge to select organisms that have received an extra gene.

Metabolism

The basis of metabolism is the same in all living cells, whether they are from barley, hops, yeast, or brewer. There is a series of reactions, generally referred to as *catabolism*, in which a feedstock is broken down to generate energy. A further series of reactions, termed *anabolism*, is responsible for building up the cellular components from which the cell is made and which are necessary for the jobs performed by the cell. This is illustrated in Figure 150 for a cell of brewing yeast.

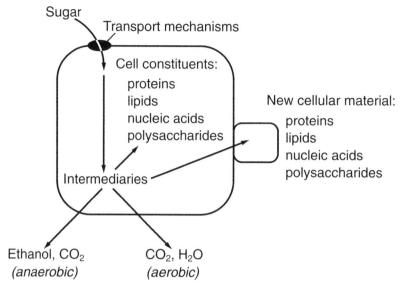

Fig. 150. Metabolism in yeast.

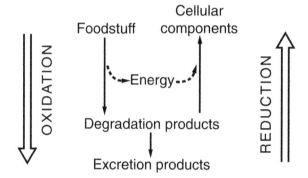

Fig. 151. Catabolism is largely oxidative and energy generating, producing intermediates that are either taken via reduction and energy consumption to the level of cellular constituents (the polymers) or are converted into excretion products (such as ethanol and carbon dioxide).

Energy is a key word in the context of metabolism. For the cells that we are primarily concerned with in the brewing industry, the main demand for energy is to fuel the biosynthetic (anabolic) reactions and to transport foodstuffs into the cells and waste products out of the cells (Fig. 151).

Energy Currency

Living cells use ATP as energy currency. In catabolic reactions that release enough energy, energy is trapped in the form of ATP. The way the cell does this is to take a molecule called adenosine diphosphate (ADP) and stick some phosphate onto it (Fig. 152). The attachment (bond) formed is a high-energy one.

When the cell needs some energy, it turns to its ATP and reverses the reaction. The phosphate is split off, ADP is reformed, and the energy is made available.

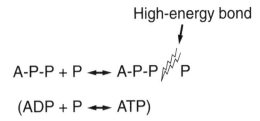

Fig. 152. Capturing energy as a bond in ATP. In the reaction headed toward the right, energy is captured (catabolic reactions). When the reaction moves to the left, energy is released and utilized (anabolic reactions).

Energy-Generating Reactions

If we want to get warm, we can burn wood. In turn, if we want to breathe life into the fireplace we pump in some oxygen. The main ingredient in wood is carbohydrate. So, taking the above at face value, if we react carbohydrate with oxygen we release energy. The smoky gas you see is primarily composed of water vapor and carbon dioxide. And so we have the famous reaction:

$$C_6H_{12}O_6 + 6\ O_2 \rightarrow 6\ CO_2 + 6\ H_2O + energy$$

Or, in words

$$Carbohydrate + oxygen \rightarrow carbon\ dioxide + water + energy$$

This is exactly the reaction that occurs in most cells that burn up sugar when they are living in the presence of oxygen. It is the principal energy-generating process in a germinating barley embryo.

In fact, the reaction doesn't happen all in one go, with an almighty release of energy, somehow captured as ATP. Rather it happens gradually and in an extremely organized way through a long sequence of reactions, each of them catalyzed by a different enzyme (see earlier for information on enzymes). The process is called *respiration*.

First, the cell needs to get the sugar into the cell. Let's take brewing yeast as our example. It makes several proteins that have the role of latching on to the various sugars and transporting them across the cell membrane. Energy is expended in achieving this.

What happens next comes as something strange to the non-biochemist: the cell takes some ATP, rips off the third phosphate, and sticks it on to the sugar. And so we have glucose-6-phosphate, or G-6-P (the phosphate is attached to carbon atom number 6 of the glucose). G-6-P is an "activated" form of glucose. It is primed for action.

There is no real point for our purposes in discussing the individual reactions (there are more than a dozen of them) that are involved in breaking down the G-6-P. We can restrict ourselves to some generalities (Fig. 153):

- The G-6-P is changed (isomerized) into a related molecule called F-6-P (F stands for fructose), which is cranked up to yet a higher energy level by the addition of another phosphate from ATP (makes F-1,6-diP).
- The F-1,6-diP, which contains six carbon atoms in the sugar portion (see earlier), is split to produce two molecules of pyruvic acid, which contains three carbon atoms.

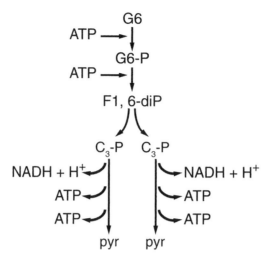

Fig. 153. Glycolysis. C_3P are three-carbon sugars energized with phosphate; pyr = pyruvic acid.

- In so doing, the cell has oxidized the sugar, with the hydrogen removed being captured by NAD (see later).
- There are two stages in the pathway where the phosphates added on to the sugar are released and recaptured as ATP. As two molecules of pyruvate are produced, then we double this value. In other words, we have expended two ATPs in activating the sugar and have produced a total of four. Net effect: capture of energy as two ATPs.

Oxidation

We considered oxidation and its opposite (reduction) earlier. For our purposes here, I will refer to oxidation as a removal of hydrogen from a molecule, whereas reduction is the addition of hydrogen.

When glucose is oxidized to pyruvate, there is a removal of four hydrogens. Where do they go? Well, just as there is a "universal" energy currency (ATP) so there are a couple of universal hydrogen carriers. The first is called nicotinamide adenine dinucleotide, henceforth to be referred to as NAD (for which you will no doubt be grateful). This tends to be the molecule that receives the hydrogens in the oxidative reactions. For sake of convenience, we will say that $NADH_2$ is produced. The second hydrogen carrier is NADP (it has an extra phosphate on it). Its role tends to be to donate hydrogen to reactions (mostly anabolic) that need hydrogen—so here we are talking about $NADPH_2$ going to NADP.

Thus, when glucose is converted to pyruvate, two molecules of $NADH_2$ are produced. If the cell merrily continued chewing glucose, then pretty soon all of the NAD would be soaked up as $NADH_2$ and no more NAD would be available. Everything would grind to a halt. So the NAD needs to be regenerated.

In cells growing in the presence of oxygen, this is achieved by transferring the hydrogen to oxygen through a series of carriers called cytochromes. Without going into details about precisely how it is done, in this transfer a total of three molecules of ATP can be made for every molecule of $NADH_2$. This is the stage at which the energy of oxidation is released and captured in a controlled way. Two molecules of $NADH_2$ need to be oxidized, a total of six ATPs at this stage.

What if the cell is growing without oxygen, though? How then does it return its $NADH_2$ to NAD? The answer is by a *fermentation* route. The $NADH_2$ is used to reduce some molecule other than oxygen. This varies between organisms, but let's concern ourselves only with *Saccharomyces*.

Yeast strips out a carbon dioxide molecule from the pyruvic acid that it has produced. This decarboxylation produces acetaldehyde, which receives the hydrogens from $NADH_2$ in a reaction catalyzed by alcohol dehydrogenase. The product, of course, is ethanol.

$$\text{Pyruvate} \rightarrow \text{acetaldehyde} + CO_2$$

$$\text{Acetaldehyde} + NADH_2 \rightarrow \text{ethanol} + NAD$$

When we break down glucose, we produce two pyruvates and two molecules of $NADH_2$. By decarboxylating and reducing both pyruvates, we can regenerate the two NADs used earlier in the pathway of glucose breakdown (known as *glycolysis*). The net effect then is

$$C_6H_{12}O_6 \rightarrow 2C_2H_5OH + 2CO_2$$

This is the equation of alcoholic fermentation by yeast—the very heart of brewery fermentations.

What Else Can Happen to the Pyruvate?

This alcoholic fermentation does not occur when yeast is growing in the presence of high concentrations of oxygen. Under these conditions, we have total conversion to carbon dioxide and water. How is this achieved? The answer is in a cyclic pathway sometimes called the Krebs cycle, sometimes the tricarboxylic acid cycle, and sometimes the citric acid cycle.

The jumping-off point demands that the pyruvate first be oxidized to acetic acid. This involves a specific enzyme (as do all the various reactions described here). In fact, free acetic acid is not produced; rather acetic acid is attached to something called coenzyme A. In just the same way that a sugar is "activated" by adding phosphate to it, so is acetic acid activated by adding it to coenzyme A, to form acetyl-CoA.

The bare bones structure (not detail) of the Krebs cycle is shown in Figure 154. This illustration deliberately does not include all the names of the intermediates, but indicates where the hydrogen and carbon dioxide emerge. Remembering that for each glucose there are two pyruvates, you can add up for yourself the emergence of six carbon dioxide molecules. And remember that every two hydrogens captured (as $NADH_2$) are oxidized to water with the release of three ATPs in the manner described earlier.

Quite clearly this complete conversion of glucose to carbon dioxide and water enables much more ATP to be produced than is captured in the anaerobic pathway. Curiously, though, if brewing yeast is confronted with a very high sugar content, it has control mechanisms that turn off the reparation route and channel the breakdown through

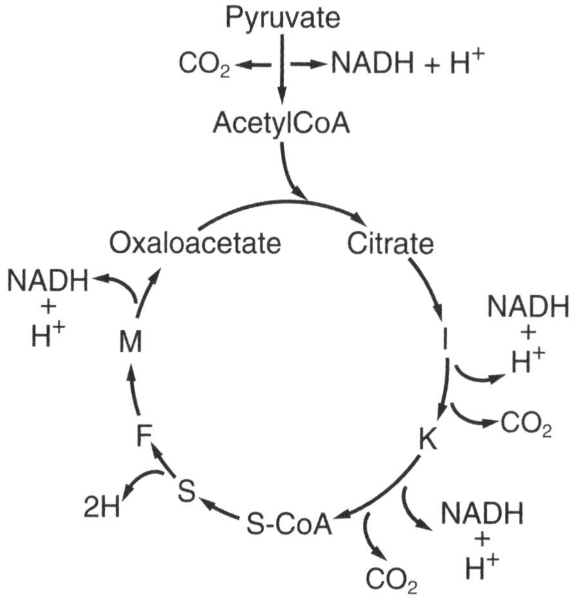

224 / Appendix 1

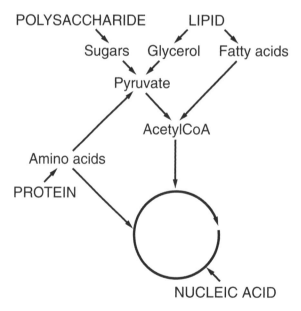

Fig. 154. Krebs cycle.

Fig. 155. Catabolic routes into the backbone of intermediary metabolism.

fermentation. This is the Crabtree effect. It seems that when confronted with all that potential energy, the cell wishes to avoid ATP overload.

What About Other Foodstuffs?

The short answer to this question is that a cell shifts the various molecules it has available for degradation to an intermediate somewhere in this central pathway (Fig. 155). If it is an endogenous store of polysaccharide (starch or glycogen, for instance) then it has enzymes that will chop the store down to glucose. Fats are converted to glycerol, which enters the pathway just above pyruvate, and acetyl-CoA, and so on.

Making Molecules

The previous section focused on how a cell such as yeast (but similar things are going on in a barley embryo or a human) degrades molecules to produce energy.

Some of that energy is used to make the molecules that the cell needs—its proteins (including the enzymes and structural proteins), the lipids for the membranes, the nucleic acids, and the storage polysaccharides (all discussed earlier). Although a cell can assimilate *some* materials preformed (for example, yeast plugs some lipids from the medium into its membranes), by and large it makes them from scratch. The molecules are built up from smaller molecules.

Without going into the specifics of the step-by-step way in which specific enzymes piece together these big molecules, note that the building bricks are intermediates in the central pathways already described (Fig. 156). It is here that the other nutrients needed by a cell come into play, e.g., sources of nitrogen and sulfur.

In other words, there are various fates for these intermediates. Let's illustrate the point by considering just one such intermediate, oxaloacetate. It is needed to capture acetyl-CoA to make citric acid to allow the Krebs pathway to cycle. However, it is also the kick off point for making a key amino acid (aspartate), which is made by sticking an amino group on to the oxaloacetate. In turn, this aspartate heads off in pathways to make more amino acids (methionine, threonine, isoleucine, and lysine) and also some of the building blocks for

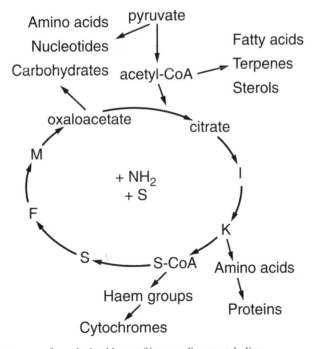

Fig. 156. Anabolic routes away from the backbone of intermediary metabolism.

DNA and RNA. Separately oxaloacetate is also the starting point for the synthesis of carbohydrates within the cell (gluconeogenesis) and many other amino acids. Some of these reactions demand reducing power, which as we have seen is delivered in the form of $NADPH_2$.

Imagine the Krebs cycle as a relay race, with each of the intermediates being runners handing on the baton in sequence. Oxaloacetate is the last of the runners, but in a position to pick up a fresh baton (acetyl-CoA) ready to start the next circuit. But if oxaloacetate has gotten sidetracked and is legging off in other directions, it needs to be replaced from another source so the race can continue. And so there are other pathways (anaplerotic) that top up the runner availability and allow the cycle to continue. They are schemes (too detailed to cover here) that effectively produce oxaloacetate from sources other than the Krebs cycle.

Exerting Control

Quite evidently, we have a somewhat complex scenario. A multitude of intercomversions is at play—it has been estimated that the very simplest of bacteria need over 1,000 separate reactions. This vast metabolic "orchestra" needs to be conducted under strict control. This is achieved by a cell in various ways, but here are just a couple.

First, the cell produces only the materials that it needs. In effect, this means that it develops only the enzymes and other workhorse molecules (e.g., transport mechanisms) needed for it to do the jobs demanded of it. To do otherwise would be a waste. Thus, if a yeast is growing on a high level of glucose, it does not make the high levels of cytochromes needed in respiration, does not make the transporters needed for other sugars present in smaller quantities, and does not turn on the enzyme needed to convert pyruvic acid into acetyl-CoA. It will, however, switch on high levels of the enzymes that convert pyruvate into acetaldehyde and then ethanol. The cell controls protein synthesis at the level of

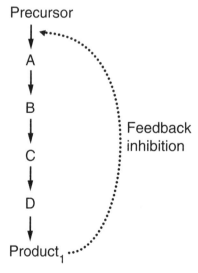

Fig. 157. Feedback control of metabolic pathways, part 1. Each of the arrows indicates a separate enzyme-catalyzed reaction. Unnecessary accumulation of the product can feed back to inhibit the enzyme catalyzing the formation of the first intermediate (A). This avoids the wasteful accumulation of the end product but also of all the intermediates in the pathway. It is likely that the precursor itself will feed back to an earlier enzyme still, and so on, in cascades designed to ensure that only necessary molecules are produced.

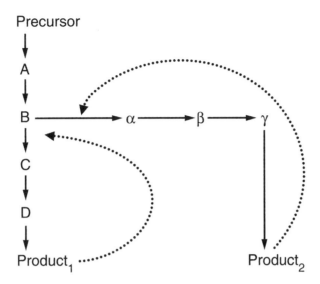

Fig. 158. Feedback control of metabolic pathways, part 2. Here if product$_1$ fed back to block the same enzyme there would be a problem, because this would prevent the formation of B, which is needed to go through the pathway to make product$_2$. Thus, in this scenario, product$_1$ and product$_2$ feed back to inhibit the first enzymes *unique* to their production.

gene transcription and translation (see earlier) and responds to triggers and hints from its environment in order to switch on or switch off the expression of various genes.

The second level of control is at the enzyme level itself and involves inhibition (see earlier). This is illustrated in Figures 157 and 158. If the end product of this pathway is accumulating to excess, it does not make sense to make any more. Then it often happens that the material "feeds back" to inhibit the first enzyme unique to the synthesis of this material. It cannot feed back too far, or it would lead to the cessation of production of something that *is* needed.

Some Necessary Chemistry

Atoms

The basic unit of all matter is the atom.

Modern thinking about the atom is all about waveforms, but it is still convenient to talk about the atom in terms of protons, neutrons, and electrons:

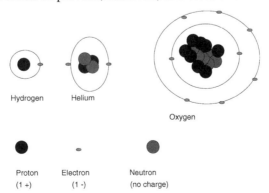

At the heart of the atom (the nucleus) are the protons and neutrons, each of which has a mass of one. Protons are positively charged, neutrons have no charge. Together they comprise the mass of the atom.

Orbiting the nucleus, rather like the planets orbit the sun, are electrons. They are negatively charged but have essentially no mass. In neutral atoms the number of electrons is exactly the same as the number of protons.

The electrons orbit the nucleus in defined orbits. There is a limit to how many electrons can occupy each orbit. The orbit nearest the nucleus can accommodate just two electrons. Because they are so close to the nucleus, the strong positive-negative interaction means that these electrons are less free to move around than those further out—i.e., they have a relatively low energy. The next orbit holds eight electrons, which, because they are that much further away, are more energetic. The next orbit holds 18, the next 32.

Each element in nature consists of atoms. There are well over 100 elements, each having successively one extra proton and therefore one extra electron. The simplest, hydrogen, has one proton and one electron. The next is helium: it has two protons (and also two neutrons, so it has a mass of 4 and not 2) and two electrons. And so on.

The most stable (least reactive elements) are those in which the outermost orbit is up to capacity with electrons (2, 8, 18, etc.). Helium, then, is very unreactive—and that is why we can be rather more comfortable flying in airships filled with helium as opposed to hydrogen.

One way in which an atom can complete its outer shell of electrons is by donating electrons to another atom, which in turn can complete is outer orbit by accepting electrons. For example, sodium (which has one electron in its outer orbital) can lose a single electron, and chlorine (which has seven electrons in its outer orbital) can gain an electron, each then assuming full outer orbits. Sodium acquires a net charge of 1+ because it has lost one electron, chloride gains a net charge of 1–, because it has one extra electron. The *compound* formed is NaCl. The sodium has become a positively charged *ion*, sometimes called a *cation,* because it is attracted to a negatively charged electrode (the cathode). Chloride has become a negatively charged ion (an *anion*).

Magnesium needs to lose two electrons, which it can by donating one electron to each of two chlorines. Thus magnesium chloride consists of one magnesium and two chlorides, $MgCl_2$. This type of bond is called an *ionic bond*.

Alternatively, an atom can complete its orbital by sharing electrons. Thus, if the orbitals of two hydrogen atoms (one electron each) come into contact with the outer orbital of an oxygen atom (six electrons), then a pair of electrons can be shared between the oxygen and each hydrogen atom to make a much more stable molecule, water, H_2O. This type of bonding is called a "covalent bond."

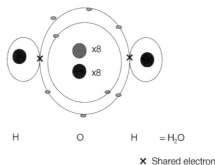

H   O   H   = H_2O

✗ Shared electron

The number of other atoms that an element can react with is known as its "valence." Thus, hydrogen reacts with only one atom at a time, valence = one; oxygen reacts with two to make it more stable, valence = two. Ergo we have water with two atoms of hydrogen and one of oxygen, H_2O. Carbon has a valence of four. Sometimes one atom is linked to another by two links—this is called a "double bond"—and it is stronger than a single bond. An atom can use up two of its valences in this way. Thus in carbon dioxide, CO_2, the carbon uses up its four valences by linking to two oxygens by double bonds, with each oxygen using up its two valences in a double bond to the carbon:

$$O=C=O$$

Oxidation and Reduction

If oxygen is added to a substance, that substance is said to be *oxidized*. Conversely, if hydrogen is added it is said to be *reduced*. Removal of hydrogen is also oxidation. If you will allow one jump of logic without elaborate explanation, when a substance loses electrons, it is also said to have been oxidized. The substance that picks up those electrons is said to be reduced.

This is one type of chemical reaction. An example is the conversion of acetaldehyde to ethanol by yeast. Acetaldehyde is reduced, and the molecule NADH, which is the substance in living organisms that carries the electrons (reducing power), has become oxidized. If one component of a system is oxidized, another component or components must be reduced. The reducing power balances.

$$\underset{\text{Acetaldehyde}}{CH_3CHO} + NADH + H^+ \leftrightarrow \underset{\text{Ethanol}}{CH_3CH_2OH} + NAD$$

Reactions

All chemical reactions also balance: the total number of atoms of a given element on the left side or a reaction must balance with that on the right. An example is the conversion of glucose to ethanol and carbon dioxide, which is the overall reaction in alcoholic fermentation of yeast:

$$C_6H_{12}O_6 \rightarrow 2\ CH_3CH_2OH + 2\ CO_2$$

A reaction happens because it is energetically favorable for it to occur. In other words, the total energy of the products is lower (more stable) than that of the reactants. The reaction may not proceed totally to the right hand side: equilibrium will be established.

This equilibrium may not be established rapidly, even if it is thermodynamically favorable. This is because bonds have to be broken in the reactants before new, more stable ones can be formed in the products. Liken it to a ball in a valley on one side of a hillock. If there were no hillock there, and next to the valley with the ball in it was another valley that was lower, the ball would tend (given a gentle nudge) to roll to the lower valley and stay there. Because of the hillock some energy has to be put in (work done) to roll the ball over the hill. But more effort would have to be exerted to move the ball in the other direction. Now if a digger were moved in to shave the top off the hillock, it would be considerably easier to move the ball from the first to the second valley (and vice-versa). But the tendency would still be for the ball to move to the valley that was lower. To return to our chemical reaction, a *catalyst* is a substance that allows chemical species to overcome the energy barrier and speed up the reaction. The catalyst is left unaltered at the end of the reaction but may have been temporarily modified during the reaction.

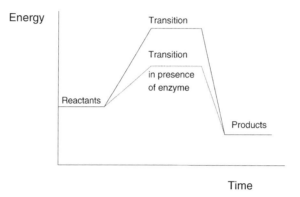

Various other factors influence the rate of chemical reactions. If the reactants are more concentrated, they have an increased opportunity to interact. If the temperature is increased, the molecules collide with greater energy and bonds are broken more readily. A good rule of thumb, first coined by Arrhenius, is that reactions occur twice as fast for every 10 degree C rise in temperature. And reactions occur much more quickly in more fluid systems. Thus, if you mix two powders together in a dry form they won't react, but if they are dissolved, the molecules can mix more freely and react together. The solvent often plays a key role in the reaction.

It is possible to group chemical compounds together into families, wherein the molecules have similar structures and similar reactivities. Three such families of great importance to the brewer are the proteins, the carbohydrates, and the lipids.

It is also possible to make sense of the complexity of chemistry by realizing that the types of "groups" that they contain determine the chemical properties of compounds. There are many types of groupings in chemistry, of which some relevant ones are

$$-C\overset{\displaystyle O}{\underset{\displaystyle OH}{\diagup\!\!\!\!\diagdown}}\quad \text{carboxyl}$$

$$-C\overset{\displaystyle O}{\underset{\displaystyle }{\diagup\!\!\!\!\diagdown}}\quad \text{carbonyl}$$

-O-H   hydroxyl

$$-N\overset{\displaystyle H}{\underset{\displaystyle H}{\diagup\!\!\!\!\diagdown}}\quad \text{amino}$$

Acids, Bases, and pH

The carboxyl group is *acidic*, in that it can furnish a hydrogen ion (H^+), i.e., a hydrogen atom without its electron, i.e., a proton.

$$-COOH \leftrightarrow -COO^- + H^+$$

The symbol $\leftrightarrow$ indicates that the reaction is reversible (can move in both directions). If the pH is low (high H^+ concentration) the reaction tends to move toward the left. If the pH is high (low H^+ concentration), the reaction will tend to move toward the right, in order to produce more H^+.

The amino group can pick up a hydrogen ion and is said to be *basic*. Another base is the hydroxide ion, OH^-.

$$-NH_2 + H^+ \leftrightarrow -NH_3^+$$

$$-OH^- + H^+ \leftrightarrow H_2O$$

The measure of acidity is the pH scale. $pH = \log 1/H^+$.

The scale runs from 1 (extremely acidic) to 14 (extremely basic), with 7.0 being neutral.

States of Matter

Matter can be basically divided into solids, liquids, and gases. They principally differ in the intensity of the forces between the component molecules and the distance between those molecules.

In a *solid* the molecules are close together, and so there is an opportunity for close-range interactions among them. The stronger these interactions, the greater the resistance of the molecules to be driven apart.

It demands energy to drive molecules apart. The most obvious source of this energy is heat, which invigorates the molecules and increases their tendency to break apart from one another. When they gain this energy and achieve mobility, they turn into *liquids*. There is still sufficient intermolecular interaction to loosely hold the molecules together. The temperature at which this solid to liquid transition occurs is, of course, the *melting point*. The stronger the links between molecules, the higher the melting point.

Put in more energy, and the intermolecular interactions are overcome, the molecules can move about largely independently—and we have a *gas*. The temperature at which the liquid-solid transition occurs is the *boiling point*. The stronger the links between molecules, the higher the boiling point.

Appendix 2

Fundamental Statistics for Brewers

The *mean* value for a set of data is the average of the values, obtained by adding them and dividing by the number of measurements.

Example
The mean of the 12 measurements 20, 18, 42, 24, 24, 24, 19, 24, 28, 21, 23, and 22 is

$$\frac{20 + 18 + 42 + 24 + 24 + 24 + 19 + 24 + 28 + 21 + 23 + 22}{12} = 24.1$$

If the individual measurements (Σ = sum of them) are x_1, x_2, x_3, etc., the total number of measurements is n, and the mean is M, then this calculation can be simplified to

$$M = \frac{1}{n} \Sigma x_i$$

The *median* is the value at the midpoint of the data set arranged in ascending order. In this case, the value is 23.5, because the data set comprises an even number of values, in which case it is conventional to take the midpoint between the middle values. Had the data set consisted of an odd number of values, the actual midpoint number would be used.

The *mode* is the most frequently occurring value in a sample set, in the present example 24.

Although there is some value to each of these measures, they tell us individually little about the spread (dispersion) of data but do give clues if we consider them together.

The *range* is the difference between the highest and lowest values, in this example 18 to 42. This tells us nothing about how the values are distributed within that range, which we can see if we plot the individual data in a histogram or a curve.

For any individual measurement, we can describe how much it deviates from the mean by the expression $x_i - M$. Summing all the deviations for an evenly distributed set of data results in a value of zero, for there are as many values higher than the mean as there are below it. Statisticians eliminate this problem by squaring the $x_i - M$ value and incorporating it into the *standard deviation* (s or σ):

$$s = \sqrt{\frac{1}{n-1} \Sigma [x_i - M]^2}$$

s^2 is called the *variance*.

Usually, for a sufficiently large data set, data adopts a "normal distribution" (Fig. 159). There is a 68.26% chance of the value being within one standard deviation of the mean, 95.44% probability of it being within two standard deviations, and a 99.73% chance of it being within three standard deviations.

$$\text{Process capability} = \frac{\text{Upper limit of measurement} - \text{lower limit of measurement}}{6\sigma}$$

Obviously, if the difference between the upper limit and lower limit is small, we have a very narrow data spread.

Brewers frequently use process control diagrams of the type shown in Figure 160. The plots incorporate reject and warning lines. The reject lines are set at three standard errors ($3\sigma/\sqrt{n}$) above and below the target. The warning lines are set at two standard errors ($2\sigma/\sqrt{n}$).

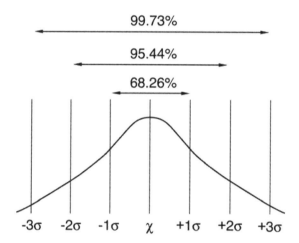

Fig. 159. Standard distribution.

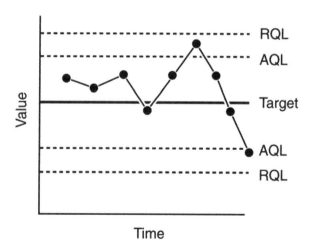

Fig. 160. Plotting production data. AQL = acceptable quality limit, RQL = reject quality limit.

A random variable (*x*) becomes *standardized* when it has been adjusted so as to have a mean of zero and a standard deviation of one. The resultant value, z, is given by $(x-\mu)/\sigma$, where μ is the *true* mean. (μ is the "theoretical" mean, as opposed to M, which is the observed mean derived from actual observations). The z value is called the *standard normal variable*.

Table 29. Normal distribution

$z = \dfrac{x-\mu}{\sigma}$	$A(z)$[a]
0.0	0.5000
0.1	0.4602
0.2	0.4207
0.3	0.3821
0.4	0.3446
0.5	0.3085
0.6	0.2743
0.7	0.2420
0.8	0.2119
0.9	0.1841
1.0	0.1587
1.1	0.1357
1.2	0.1151
1.3	0.0968
1.4	0.0808
1.5	0.0668
1.6	0.0548
1.7	0.0446
1.8	0.0359
1.9	0.0287
2.0	0.0228
2.1	0.0179
2.2	0.0139
2.3	0.0107
2.4	0.0082
2.5	0.0062
2.6	0.0047
2.7	0.0035
2.8	0.0026
2.9	0.0019
3.0	0.00135
3.1	0.00097
3.2	0.00069
3.3	0.00048
3.4	0.00034
3.5	0.00023
3.6	0.00016
3.7	0.00011
3.8	0.00007
3.9	0.00005
4.0	0.00003

[a] $A(z)$ is the area under the standardized normal plot. Notice that the uppermost value is 0.50. This is when $z = 0$, i.e., when z is actually the true mean (center point, or peak, of the plot). Obviously, for this value, half of the plot (0.5, or 50%) is to the left of this value and half to the right.

These values allow us to calculate the extent to which samples that we are interested in are *likely* to be outside certain analytical limits. Tables are available giving values that allow one to work out normal probabilities of this (Table 29).

Example

The target DMS level in packaged beer is 40 ± 5 ppb, i.e., the lower limit is 35 and the upper limit is 45. Routine analysis over time has shown that the measured DMS shows a normal distribution about a mean of 35.8, with a standard deviation of 7.68. What proportion of beer batches are likely to be outside the specified limits?

For the lower limit, $x = 35$, and therefore

$$z = \frac{(35 - 35.8)}{7.68} = -0.1$$

By reference to the table we see that the probability level is 0.46. Thus, there is a 46% chance of the DMS value being below the targeted minimum. Similarly, for the upper level we can calculate that

$$z = \frac{(45 - 35.8)}{7.68} = -1.19$$

Again referring to the table, we find that the probability level of a sample being too high in DMS is around 0.115, i.e., 11%.

It seems that the brewer in this example needs to take measures to increase its DMS levels.

Index

Abscisic acid, 35
Absorbency, 179
Acetaldehyde, 132, 229
Acetic acid, 122, 171, 206
Acetobacter spp., 171
Acetoxylan esterase, 47
Acetyl-coenzyme A, 127
Acetyl transferase (AAT), 127
Acid-washing, 109
Acidification power, 110
Acidity, 231
α-Acids (*see also* Resins), 85, 86–87, 88–89
Acrospires, 38
Adenine, 214–217
Adhumulone, 87, 89
Adjuncts, 4, 68–69
Adlupulone, 87
ADP (adenosine diphosphate), 220–222
Adsorption, 82
Aeration, 100
Aerobacter spp., 171
African beers, 11
Aging, 134
Agitation, 141
Agronomics, 30–31
Air-off temperature, 40
Air rest, 29, 33
Air-on temperature, 40
Albumins, 25, 54
Alcohol chilling test, 146
Alcohol content, 1, 2, 10, 124, 178–179
Alcohol dehydrogenase, 81, 157
Alcohols, 124–126, 127, 153
Aldehydes, 153
Aldol condensations, 153
Ales
    color and, 10
    conditioning of, 7
    fermentation and, 6, 114
    Saccharomyces cerevisiae and, 5
    sulfur compounds and, 129
    traditional, 9
Aleurone, 22, 27, 35, 49

Alginate, 141, 164, 206
Alkalinity, 81, 147
All-malt worts, 104
American Malting Barley Association (AMBA), 26, 27
Amino acids, 106, 126, 153, 164, 189–193
Amino groups, 190, 191, 231
Ammonium sulfate, 181
Amphipathicity, 162, 211
α-Amylase, 52
β-Amylase, 164
Amylases, 52–54, 73, 81, 162–164, 194
Amylopectin, 50, 52, 54, 203
Amylose, 52, 203
Anabolic fermentation, 125
Anabolism, 125, 219–220, 225
Anaerobic fermentation, 5
Analysis
    aging and, 159
    of alcohol content, 178–179
    of bitterness, 181
    brew house operations and, 178
    of carbon dioxide content, 179
    of clarity, 180
    of colloidal shelf life, 146
    of color, 179
    of contents, 179
    of diacetyl, 181
    of dissolved oxygen, 180
    of extracts, 179
    flavor and, 181–182
    of foam, 165–166, 182
    gushing and, 168
    homogeneity and, 49
    of hops, 85, 86
    of metal content, 182
    microbiological, 182–183
    of β-glucans, 46
    of pH, 179
    sensory, 183–184
    stability predictions and, 180–181
    of yeast strains, 101

Aneuploidy, 103
Animal feeds, 186, 187
Anions, 228
Annealing, 51
Antifoam, 116
Antioxidants
    beer as source of, 18
    polyphenols and, 45
    specialty malts and, 42
    stabilization and, 137
    staling and, 156–157, 159
Aphids, 83
Apparent total nitroso compounds (ATNCs), 19
Apple flavor, 126, 132
Apples, 3
Arabinofuranosidase, 49
Arabinoxylan (*see* Pentosans)
Aroma
    α-acids and, 88
    alcohols and, 124–126
    essential oils and, 90
    esters and, 126–127
    flavor and, 16–17
    hops and, 131
    hops extracts and, 93
    malt and, 131–132
    overview of, 123–124, 132–133
    sulfur compounds and, 128–131
    vicinal diketones and, 127–128
Aroma hops, 85, 86, 147
Artificial noses, 182
Ascorbic acid, 159
Aspergillus spp., 31
Astringency, 13, 81
Atoms, 227–229
ATP (adenosine triphosphate), 199, 220–222
ATP bioluminescence, 182–183
Attenuation, 7, 114
Autolysis, 114
Autoxidation, 155, 156
Auxiliary finings, 135

Bacteria, 141, 171–172, 177, 182–183
Banana flavor, 16, 126
Barley
　as adjunct, 68
　agronomics of, 30–31
　cell wall polysaccharides of, 45–49
　commercial malting stages of, 22–23
　composition of, 50
　foam and, 164
　husks of, 3
　kernel size and, 28
　lipids of, 55–57
　lipoxygenase and, 154
　S-methylmethionine and, 129
　modification potential of, 28–29
　nitrogen content of, 27–28
　plant structure and, 21–25
　polyphenols and, 57, 147
　proteins of, 54–55
　sampling of, 30
　selection of, 7
　starches of, 4, 49–54
　unmalted, 21
　varieties of, 26–27
Barley brewing, 21
Barley factors, 27
Barley reception, 31–32
Barms, 110
Bed porosity, 69
Bicarbonate, 81
Biological oxygen demand, 187
Bioluminescence, 182–183
Birefringence, 50
Bisulfite, 152, 156–157, 158, 159
Bits, 141
Bitter ales, 10
Bittering hops, 85, 86–87, 196–197
Bitterness
　α-acids and, 88, 89, 122, 123
　analysis of, 181
　flavor and, 16, 151
　iso-α-acids and, 5
　pre-isomerized extracts and, 91
　resins and, 85
Black malt, 43
Body, 13, 16, 82, 121, 133
Body feeds, 137
Boiling
　additives and, 97–98
　brew kettles and, 95–97
　colloidal shelf life and, 148
　defined, 7
　essential oils and, 90
　foam and, 162
　function of, 4–5, 95
　hardness and, 82
　temperature and, 1
Boiling points, 231
Bonds
　covalent, 228
　glycosidic, 203
　hydrogen, 192–193, 209
　ionic, 228
　peptide, 191
Bottles
　filling of, 174–175
　gushing and, 167, 168
　light instability and, 169
　waste and, 187, 188
Bottom fermentation, 102, 164
Bracteoles, 84, 85
Break point, 40
Breeding, 26, 27
Brettanomyces spp., 172
Brew house operations, 72–75, 178
Brewing yeasts (*see* Yeast)
Bright beer, 137, 144
Brightness, 69
Brine, 99
Bromate, 34, 157
Brown ales, 10
Browning reaction, 17, 95, 132, 203
Budding, 103
Buffers, 191
Burton malts, 80
Burton union system, 116
Butterscotch flavor, 6–7, 115, 127

Cakes, 110
Calandrias, 96, 97
Calcium
　boiling and, 97
　brew house operations and, 74
　gushing and, 167
　mashing and, 4
　oxalate and, 147
　role of, 81
　temporary hardness and, 82
　visible hazes and, 141
Calcofluor, 43, 46, 49
Calculations, 75–76
Candida spp., 172
Cannibis sativa, 83
Canning, 175
Capacitance probes, 110

Cara Pils, 43
Caramels, 17, 43
Carbohydrates
　chemistry of, 199–206
　energy and, 221–222
　as food stores, 198–199
　groups sticking to, 205–206
　structural role of, 199
　sweet wort and, 73
　visible hazes and, 141
　yeast metabolism and, 104–105, 107
Carbon dioxide
　analysis of, 179
　flavor and, 5
　foam and, 17, 182
　mouthfeel and, 13, 16
　stabilization and, 138
　units for, 2
Carbonation, 115, 138, 167
Carbonyls, 152, 153
Carboxyl groups, 190, 191, 206–209, 231
Carboxypeptidase, 47, 55
Carcinogens, 82
Cardiovascular disease, 18
Carica papaya, 146
Carrageenan, 97, 148
Caryophyllene episulfide, 131
Catabolic fermentation, 125, 126
Catabolism, 219–220, 224
Catalysts, 194, 230
Catechin, 56, 57, 143
Cations, 228
Cattle feed, 186, 187
Cell number, 110
Cell wall components, 45–49, 72–73
Cellobiose, 24, 203
Centrifugation, 135
Cereal, 68
Chapon model, 143, 146
Cheesy character, 89
Chelation, 157
Chit malts, 42
Chitting, 28, 33
Chloride, 14, 82, 122
Chocolate malt, 43
Chromaticity, 179
Cider, 3
Citric acid cycle, 223–224
Citrobacter spp., 171
Clarification, 7, 98–99, 134–135, 135–137
Clarity, 17, 148, 180
Clark (*see* Ross and Clark method)
Classification of beers, 9–11

Cleaning-in-place (CIP) systems, 117–118, 182
Cleaning water, 80
Cleanings, 175
Cling, 165
Cloning, 218
Clove-like character, 132
Coagulants, 148
Codons, 217
Cohumulone, 87, 89
Cold break
  cold conditioning and, 134
  cooling and, 95, 99–100
  cylindroconical vessels and, 116
  defined, 95
  pitching and, 5
  removal of, 113
Cold conditioning, 115, 134–135, 148
Cold storage, 7
Cold water extract, 43
Collagen, 135
Colloidal instability, 141
Colloidal shelf life, 146, 148
Color
  analysis of, 179
  beer classification and, 10
  boiling and, 95
  caramels and, 17, 43
  polyphenols and, 147
  quality and, 17
  specifications for, 44
  Vienna malts and, 42
Colupulone, 87
Commercial malting, 22–23
Competitive inhibitors, 196
Concentrations, 230
Condensations, 153, 205
Conditioning, 1, 7, 115
Conductance test, 86
Cone-to-cone pitching, 116
Cones, 84, 85, 90, 99
Congress wort, 27
Contaminants
  analysis of, 182–183
  aroma and, 132
  bacteria as, 171–172
  continuous fermentation and, 119
  diacetyl formation and, 128
  gushing and, 167
  stabilization and, 137
  wild yeast as, 172
Continuous fermentation, 118–119
Conversions, 2
Cooling, 99–100, 164

Copper, 81, 155, 195
Copper vessels, 136–137, 158
Corn, 68
Corn sugars, 97
Costs, 1, 68, 188
Covalent bonds, 228
Crabtree effect, 109
Crystal malt, 43, 132, 164
Cultivation, 30–31, 83–84
Curing, 40
Cylindroconical vessels, 116–117
Cytoduction, 104
Cytosine, 214–217
Czech Republic, 42

Daltons, 191
Damson-hop aphids, 83
Darcy's equation, 69
De Vries equation, 161
Deaeration, 82
Debaromyces spp., 172
Decoction mashing, 66, 67
Degradation, 88
Demineralization, 82
Deoxynivalenol, 19, 167
Descriptive tests, 184
Destoners, 60
Detergents, 17, 118, 162, 163
Deviation, standard, 233
Dextrinase (*see* Limit dextrinase)
Dextrins, 104, 121
Diacetyl (*see also* Vicinal diketone), 6–7, 16, 134, 171, 181
Diastatic malts, 42
Diatomaceous earth (*see* Kieselguhr)
Difference tests, 183
Dimethyl sulfide
  aroma and, 16, 131
  boiling and, 95
  flavor and, 124, 129
  hops and, 131
  lager malts and, 41
  sources of, 130
Dipeptides, 191
Disaccharides, 203
Discrete polypeptide hypothesis, 162
Diseases, 83
Disproportionation, 161
Dissolved oxygen, 180
Distributions, 234
Disulfide bridges, 193
Diterpenes, 89
DNA (deoxyribonucleic acid), 27, 104, 214–219

Dominion Breweries, 118
Dormancy, 28
Double helixes, 214, 215–217
Downy mildew, 83
Draft beers, 11
Drainage, 161
Dry beers, 11
Dry hop character, 90
Dry hopping, 5
Dry milling, 59–62, 63
Drying (*see also* Kilning), 31, 81
Dual-purpose hops, 85
Dwarf hops varieties, 83

Ehrlich fermentation, 125, 126
Electrodes, 179, 180, 181
Electron spin resonance (ESR) analysis, 159
Electrons, 227–229
Embden-Meyerhof-Parnas (EMP) glycolysis pathway, 105
Embryos, 22
Endoenzymes, 47, 52
Endogenous antioxidants, 156–157
Endopeptidases, 55, 146, 164, 194
Endosperm
  germination and, 4
  β-glucan and, 204–205
  invisible hazes and, 141
  kernel structure and, 22, 23
  malting barleys and, 7
  modification of, 23
  texture of, 29
Energy
  carbohydrates as, 199
  catabolism and, 219–220
  chemical reactions and, 229–230
  currency of, 220–221
  lipids and, 213–214
  reactions generating, 221–222
Energy consumption, 185, 186
Enterobacteriaceae, 171, 172
Environmental impacts of brew house operations, 185–188
Enzyme-substrate complexes, 195
Enzymes
  aleurone layer and, 22
  breeding guidelines for, 27
  calcium ions and, 81
  diastatic malts and, 42
  feedback control and, 226–227

foam and, 146
germination and, 4
β-glucans and, 46–48
green malts and, 42
haze removal and, 146
hormones and, 35
importance of to brewers, 194–196
mash thickness and, 74
nitrogen content and, 28
polysaccharides and, 205
specifications for, 44
staling and, 154, 157
zinc and, 81
Epicatechin, 57
Equilibrium, 229–230
Ergosterol, 210
Escherichia spp., 171
Essential oils, 5, 16, 89–90, 93
Esters, 47, 125, 126–127, 206, 209
Etching, 165
Ethanol, 17, 18, 124–125, 164
Ethyl acetate, 126
Ethyl caprylate, 126
European Brewing Convention Units, 142
Excise duty, 1
Exoenzymes, 52
Explosions, 63
External wort boilers, 96
Extracts, 43, 90, 91–92

Farbebier, 43
Farinators, 32
Farming, 30–31
Fats, 209
Fatty acids
α-acids and, 89
amphipathicity of, 212
in barley, 55
flavor and, 132
lipid structure and, 206–209, 210–211
oxygen and, 106
staling and, 154
Fed-batch culturing, 109
Feed barleys, 26
Feedback control, 226–227
Fermentability, 44
Fermentation
alcohols and, 125
ales and, 9
cleaning-in-place systems and, 117–118
cold break and, 99–100
contaminants and, 171
continuous, 118–119

control of, 113–114
defined, 7
dimethyl sulfide and, 130
energy and, 223–224
foam and, 164
high-gravity brewing and, 118
secondary, 115
staling and, 158
temperature and, 1
time of, 6
waste and, 187
by yeast, 3
zinc and, 81
Fermenters, 115–117
Ferulic acid, 49, 132
Feruloyl esterase, 47
Filling operations, 174–175
Filter beds, 3, 22, 71–72, 133
Filtration
clarification and, 135–137
foam and, 164
materials for, 3, 22, 71–72, 133
overview of, 7
sterile, 173
waste and, 187
as water treatment, 82
Fine/coarse difference, 43
Finings
bit formation and, 141
boiling and, 97
chemistry of, 206
cold conditioning and, 135
sedimentation and, 148
traditional ale brewing and, 7
Fixing, 95
Flaking, 68, 69
Flash pasteurization, 173
Flavanols, 147
Flavor
aging and, 134
analysis, 181–182
aroma and, 16–17
assessment of, 183–184
bitterness and, 16
changes in, 151–153
continuous fermentation and, 119
diacetyl and pentanedione and, 7
fermentation vessels and, 117
iron and, 81
mouthfeel and, 133
Munich malts and, 41
oxidation and, 137
quality and, 13–17
salt and, 14

smell and, 123–133
sourness and, 13–14
specialty malts and, 42–43
specifications for, 44
sweetness and, 13
taste and, 121–123
tongue and, 13, 15
Flavor instability, 141
Flavor threshold, 123–124
Flavor units, 124
Flavor wheel, 13, 14
Flocculation, 81, 114
Flour, 68
Fluorescein dibutyrate staining, 28
Foam
analysis of, 182
assessment of, 165–166
components of, 162–164
gushing and, 167–168
high-gravity brewing and, 118
lipids and, 211–213
nitrogen gas and, 16
physics of, 161–162
polypeptides and, 162
process effects on, 164–165
quality and, 17
stabilization and, 141, 146, 148
zinc and, 81
Foam instability, 141
Folate, 18
Folding, 192
Food stores, carbohydrates as, 198–199
Formaldehyde, 167
Fractionation, 90
Free amino nitrogen (FAN), 126
Free drying, 40
Freezing, 108
Friability, 4, 43
Fruity flavor, 126, 151
Fullness, 82
Fungi, 167
Fusarium sp., 19, 31, 167

α-Galactosidase, 101
Gas chromatography, 179, 181
Gas control, 138
Gases, 167, 231
Gelatinization, 50–52, 68, 141
Gels, 46
Generalized amphipathic hypothesis, 162
Genetic code, 215–217
Genetic modification, 217–219
Genotypes, 101

Geraniol, 90
Germination
　defined, 7
　β-glucan degradation and, 48
　malting process and, 23
　pentosan breakdown and, 49
　process of, 4, 35–39
　protein degradation and, 54–55
　temperature and, 1
Germinative capacity, 28, 29, 109–110, 113
Gibberellic acid (GA), 34–35, 42, 55
Glass, 167, 168, 169
Globulins, 54
β-Glucanases, 7, 46, 47–48, 72
Glucanolytic standards, 1
α-Glucans, 141
β-Glucans
　as barley cell wall component, 45–46
　breakdown of, 46–49
　endosperm and, 204–205
　germination and, 36
　modification evaluation and, 43
　starchy endosperm structure and, 23–24
　visible hazes and, 141
　yeast and, 102, 103, 206
　yield and, 45
Glucans, defined, 201
Glucoamylases, 54, 73
Gluconeogenesis, 225–226
Gluconobacter spp., 171
Glucosans, 201
Glucose, 198, 199, 202, 222–223
β-Glucosidases, 105
Glutelins, 54
β-Glycanases, 194
Glyceraldehyde, 199–201
Glycerides, 208, 209
Glycerol, 208–209
Glycogen, 107
Glycolipids, 209
Glycolysis, 105, 221–222
Glycosidic bonds, 203
Granule annealing, 51
Grapes, 3
Gravity (*see also* Specific gravity), 135
Gravity meters, 179
Green malts, 42
Grist, 11
Grist cases, 64
Grits, 68, 69
Growing season, 30–31

Guanine, 214–217
Gum, 24
Gushing, 17, 141, 167–168

Hammer mills, 62, 63
Hansen, Emil Christian, 107
Hansenula spp., 172
Hard cider, 3, 144
Hardness, 77, 80, 82
Haze, 141–144
　gushing and, 167
　holistic stabilization of, 146–148
　instability of, 141
　iron and, 81
　measurement of, 142, 180
　polyphenols and, 17, 45, 57
　predictive methods for, 146
　protein interactions and, 197
　silicate and, 82
　stabilization and, 137, 144–146
　tannoids and, 88, 137
Haze-active polyphenols, 142–144
Haze-active proteins, 142
Haze instability, 141
Haze meters, 180
Head, 133
Headspace, 174
Health, 18–20, 82, 137
Heat exchangers, 99
Heavy metals, 19, 167
Hemicellulose, 24, 186
Hemocytometers, 110
Heterofermentative bacteria, 171
High-gravity brewing, 118
High-performance liquid chromatography (HPLC), 181
Holistic approaches, 147–148
Homofermentative bacteria, 171
Homogeneity, 49
Honey flavor, 16, 127
Hop back, 98
Hop oils, 5, 16, 89–90, 93
Hoppers, 64
Hoppy noses, 5
Hops
　α-acids and, 123
　active ingredients of, 84
　analysis of, 85
　aroma and, 131
　chemistry of, 87–90
　components of, 5
　cultivation of, 83–84
　growth process of, 84–85
　high-gravity brewing and, 118

　polyphenols and, 142
　preparations of, 90–93
　types of, 85–87
　units for, 2
Hordeins, 25, 54, 142, 162, 164
Hordeum distichon, 21
Hordeum vulgare, 21
Hormones, 35, 35–36
Hot break, 5, 187
Hot/cold cycling test, 146
Humulene, 89
Humulus japonicus, 83
Humulus lupulus, 83
Husks, 3, 22, 186
Hybridization, 104
Hydration, 4, 7, 22, 32–35
Hydrogels, 145
Hydrogen bonds, 192–193, 209
Hydrogen peroxide, 28, 156
Hydrogen sulfide, 129, 130–131, 169
Hydrolysis, 36, 46–47
Hydrometers, 179
Hydroperoxide, 154
Hydrophilicity, 193, 211
Hydrophobicity, 162, 164, 167, 193, 211
Hydrophobin, 167
Hygiene, 182–183
Hypochlorite, 35
Hypoxanthine, 217

Ice beers, 11
Imino acids, 189
Immobilized yeast, 119
Inactivators, 195
Incomplete charges, 209
Indirect firing, 39–40
Infection (*see* Contaminants)
Infusion mashing, 65–66
Inhibition, 195–196, 226–227
Inositol phosphate, 74
Insolubility, 209–211
Interrupted steeping, 29, 33
Invertase, 104
Inverted sugars, 97
Invisible hazes, 141, 180
Iodine test, 52
Ionic bonds, 228
Ions
　chemistry of, 228
　foam and, 162
　protein interactions and, 196–197
　role of key individual, 81–82
　saltiness and, 14
　sourness and, 13
　staling and, 155, 157

Iron, 14, 81, 133, 155, 181
Isinglass finings, 135, 141, 162, 189
ISO 9000, 177
Iso-α-acids
  α-acid isomerization and, 88, 89
  bitterness and, 5, 16, 122, 181
  contaminants and, 171
  foam and, 17, 162, 163
  light instability and, 169–170
  polypeptides and, 196–197
  staling and, 153
  sulfur compounds and, 131
Iso-amyl acetate, 16, 126
Iso-amyl alcohol, 16
Isoadhumulone, 164
Isohumulone, 164
Isomerization
  of α-acids, 88, 89, 92, 95, 123
  boiling and, 5, 95
  hops preparations and, 91–92
Isomers, 200
Isovaleraldehyde, 132

Jackets, 96, 97

Karahenone, 89
Kegging, 175
Kernels, 22, 28
Kettles, 4–5, 136–137, 158
Kieselguhr, 136–137, 187
Kilning
  color and, 17
  defined, 7
  lipoxygenases and, 56
  malts and, 4, 23
  process overview, 39–41
  stages of, 40–41
  temperature and, 1, 39
Kilns, 39, 42
Klebsiella spp., 171
Kloeckera spp., 172
Knockout, 98
Krausen, 115
Krausening yeast, 128
Krebs cycle, 223–224

Laccase, 57
Lactic acid, 122, 171
Lactobacillus bacteria, 74, 128, 171
Lagering, 41, 115, 134
Lagers
  cold storage and, 7
  color and, 10
  conditioning of, 7
  dimethyl sulfide and, 16
  fermentation and, 6, 114
  Saccharomyces pastorianis and, 5
  sulfur compounds and, 129
  traditional, 9
Lambic/gueuze products, 14
Laminaribiose, 24
Late hop aroma, 16, 90, 131
Late hopping, 5
Lauter tuns, 66, 70–71
Lead, 19
Lead conductance test, 86
Life cycle, yeast, 103–104
Light, 91–92, 169–170
Light beers, 11
Light instability, 141
Light scatter, 110, 180
Light-struck, 169
Lightly kilned malts, 42
Lignocellulose, 186
Limit dextrinase, 52, 54, 73
Linalool, 89, 90
Linoleic acid, 55, 154, 208
Linolenic acid, 55
Lipases, 56
Lipid-binding proteins, 162
Lipid transfer protein (LTP1), 162
Lipids
  amphipathicity of, 211
  in barley, 55–57
  energy and, 213–214
  fats and oils and, 209
  foam and, 17, 162, 163, 211–213
  insolubility of, 209–211
  oxygen and, 213
  structure of, 206–209
  yeast and, 102, 104
Lipoxygenases, 56–57, 154, 156, 157, 158
Liquids, 231
Liquor, 65
Lovibond Tintometers, 179
Low-alcohol beer, 10
LOX, 154
Lubricants, 141
Luciferase, 182–183
Lupulone, 87

Magnetic separators, 60
Maillard reaction, 17, 95, 132, 203
Malt factors, 27
Malternative beers, 11
Maltese cross pattern, 51
Malting
  barley reception and, 31–32
  essentials of, 7
  foam and, 164
  germination and, 35–39
  kilning and, 39–41
  overview of malt types and, 41–44
  process overview, 4
  stages of, 7
  steeping and, 32–35
Malting loss, 22
Malting premiums, 27
Malting varieties, 26, 41–44, 46
Maltose, 104, 105, 121, 203
Maltotriose, 104, 105
Malts, 41–43, 43–44, 68, 131–132, 142
Mannoprotein, 102, 103
Mannose, 202
Mash filters, 71–72
Mash mixers, 67
Mashing
  brew house operations and, 73–74
  decoction, 66, 67
  defined, 7, 64
  foam and, 164
  infusion, 65–66
  pH and, 74
  process overview, 4
  processes for, 64–65
  temperature and, 47–48
  temperature-programmed, 66–68
Matter, states of, 231
Maturation, 7
MBT, 91–92, 131, 169
Mealiness, 29, 32, 36
Mean, 233
Measurements, 142
Mechanical germination plants, 36–38
Median, 233
Megasphaera spp., 171, 172
MEL gene, 101
Melanoidins
  color and, 17
  foam and, 162, 164
  Munich malts and, 41
  staling and, 153, 156, 157
Melibiose, 101
Melting points, 231
Metabolism
  controlling, 226–227
  energy and, 220–222
  feedback control of, 226–227
  molecule creation and, 225–226
  overview of, 219–220

oxidation and, 222–223
pyruvate and, 223–224
Metallic character, 133, 182
Methanethiol, 169
Methional, 169
Methylene blue, 110
Methylene violet, 110
Micelles, 211
Microbubbles, 167
Migration tendencies, 102
Mild ales, 10
Mildew, 83
Milling, 4, 7, 59, 59–62, 63
Mineral wool, 115
Minerals, 18, 104
Mitochondria, 103
Mode, 233
Modification (*see also* Genetic modification), 23, 27, 35–36, 42, 43
Modification potential, 28–29
Mogden formula, 188
Moisture content (*see also* Steeping)
  of barley, 28
  germination and, 32
  kilning and, 23
  malt types and, 41–42
  specifications for, 43
  steeping and, 7, 22
Molds, 83, 84, 167
Molecule assembly, 225–226
Monochloropropanols, 19
Monosaccharides, 203
Monoterpenes, 89
Mouthfeel, 13, 16, 82, 121, 133
Munich malts, 41–42
Mutation, 104, 217–218
Mycotoxins, 19, 167
Myrcene, 90, 131

NAD (nicotinamide adenine dinucleotide), 222–223
Near-infrared (NIR) spectroscopy, 179
Nephelometric turbidity units, 142
Neutrons, 227–229
NIBEM method, 165–166, 182
Nickel, 167
Nitrates, 20, 82
Nitrogen, 104, 105
Nitrogen content (*see also* Free amino nitrogen (FAN))
  of barley, 27–28
  foam and, 164
  Munich malts and, 41
  pale malts and, 42

six-row barleys and, 26
specifications for, 43
storage proteins and, 54
Nitrogen gas, 16, 165
Nitrokegs, 11, 138
Nitrosamines, 19, 40, 44, 82, 172
Nitrosodimethylamine, 40
Nonbiological instability, 141
Noncompetitive inhibitors, 196
Nonenal, 152
Normal distribution, 234
Nucleation, 17, 167
Nuclei, 228–229
Nucleic acids, 214–219

Oberteig layers, 56–57
Obesumbacterium spp., 171
Odors, 187
Oils (*see also* Essential oils), 84, 209
Oleic acid, 55, 208
Oligopeptides, 191
Oligosaccharides, 203
Open tank fermentation systems, 116
Orbits, electron, 228–229
Original extract, 9–10, 179
Original gravity, 9–10, 179
Osmotic pressure, 198
Oxalate, 81, 141, 147
Oxaloacetate, 225–226
Oxidation
  flavor and, 137
  iron and, 81
  lipids and, 213
  metabolism and, 222–223
  overview of, 229–231
  protein-polyphenol haze and, 142
  staling and, 153, 154–155
Oxygen
  analysis of, 180
  fermentation and, 114
  haze-active polyphenols and, 143
  high-gravity brewing and, 118
  lipids and, 213
  polyphenols and, 147
  staling and, 155–158
  steeping and, 33
  yeast metabolism and, 106–107
Oxygen consumption, 110
Oxygen content, 82, 137

Packaging, 7, 90, 148, 173
Pale ales, 10

Pale malts, 42
Palmitic acid, 55, 210
Papain, 146
Papery character, 151
Paraflows, 99
Parti-gyling, 71
Pasteurization
  foam and, 164
  gushing and, 167
  oxygen measurement and, 180
  packaging and, 173
  recovered beer and, 139
  temperature and, 1
Pasteurization unit (PU), 173
Peat, 42
Pectinatus spp., 171, 172
Pediococcus bacteria, 128, 171
Pellets, 90, 91
Penicillium spp., 31
Pentanedione, 7, 127
Pentosans
  as cell wall component, 46, 49, 205
  defined, 201
  starchy endosperm structure and, 23, 24–25
  visible hazes and, 141
  yield and, 45
Peptide bonds, 191
Percent evaporation, 95
Perlite, 136, 137
Permanent hardness, 77
Permeability, 69
Permeases, 105, 106
Peroxidase, 57, 147, 156
Pesticides, 19, 31–32
Pests, 31, 83
pH
  amino acid charge and, 190–191
  analysis of, 179
  bicarbonate and, 81
  boiling and, 97
  brew house operations and, 74
  calcium ions and, 81
  enzyme function and, 195
  fermentation and, 114
  finings and, 135
  foam and, 164
  overview of, 231
  phosphate and, 82
  staling and, 154
  taste and, 122
Phenolic character, 115, 132
Phenotypes, 101
Phenylethanol, 16, 130

Phenylethyl acetate, 16, 126, 130
Phosphates, 74, 82, 205–206
Phospholipids, 209
Phosphomannan, 205
Phosphoric acid, 109, 214, 215
Phytate, 74
Pichia spp., 172
Pilsner malts, 11, 41, 80
Pitching, 5, 110, 113, 118
Plant structure, 21–25, 199
Plasmids, 218
Plastics, 159
Ploidy, 219
Pneumatic germination plants, 36–38
Polymerization, 142
Polypeptides, 162, 164, 169, 191, 196–197
Polyphenols (*see also* Laccase)
  in barley, 57
  color and, 17
  examples of, 56
  function of, 45
  haze and, 17, 137, 141–144, 180–181
  interactions of with proteins, 197
  removal methods for, 144–146
  staling and, 153, 156, 157
Polypropylene, 71
Polysaccharides, 102, 203, 205
Polyvinyl-polypyrollidone (PVPP), 142, 143, 144–145
Porosity, 69
Potable water, 77, 78–79
Potassium, 14, 122
Potassium bromate, 34, 157
Powder filtration, 135
Pre-isomerized extracts, 43, 90, 91–92
Pre-mashers, 65
Precipitation, 162
Precipitation reactions, 82
Precipitation tests, 146
Precoating, 137
Preevacuation, 159
Pregermination, 28, 31
Premashers, 158
Preparations, 90–93
Presses, 138–139
Primary fermentation, 113–114
Primary gushing, 167
Primary structure, 192
Procyanidin, 56
Prodelphinidin, 56
Product inhibition, 196
Profiling, 184

Proline, 142, 189
Propagation, 108–109
Propylene glycol, 99
Propylene glycol alginate, 141, 164, 206
Proteases, 55, 146, 164, 194
Protein middle lamella, 46
Protein-polyphenol hazes, 142
Protein rest, 73
Protein sensitivity, 146
Protein Z, 162, 164
Proteinases, 146, 164, 194
Proteins
  amino acids and, 189–193
  analysis of for varietal determination, 27
  in barley, 54–55
  boiling and, 95, 97
  brew house operations and, 73
  foam and, 17, 162, 163
  germination and, 36
  haze and, 137–138, 142, 180
  importance of to brewers, 194–196
  interactions of with other molecules, 196–197
  interactions of with polyphenols, 197
  nitrogen content and, 27–28, 44
  peptide bonds and, 191
  solubility and, 197–198
  stabilization and, 144
  starchy endosperm structure and, 23, 25
  translation and, 217
  visible hazes and, 141
  yeast structure and, 102
Proteolytic stands, 147
Protons, 227–229
Pseudo hazes, 141, 180
Pseudoperonospora humuli, 84
Purindolenes, 162
Pyruvate, 223–224

Quality, 13–17, 18–20, 68
Quality assurance, 177, 182
Quality control, 177
Quercetin, 56

Rachilla, 27
Radicals, 155, 156, 157, 159
Range, 233
Rare mating, 104
Rauch-bier, 42
Real extract, 179
Reception, 31–32

Recombinant DNA technology, 104
Recovered beer, 138–139
Reducing groups, 203
Reductases, 128
Reduction, 91, 92, 123, 157, 229–231
Refractometry, 179
Regenerable PVPP, 144–145
Reinheitsgebot, 42, 43, 74
Repeatability (r), 178
Reproducibility (R), 178
Residual beer, 187
Residual extract, 179
Resins, 5, 84, 87–89, 123
Resource consumption, 185–188
Respiration, 221–222
Retrograded starch, 52
Rhodotorula spp., 172
Ribes character, 151
Riboflavin, 169
Ribose, 199, 214, 215
RNA (ribonucleic acid), 199, 214–219
Roll mills, 61–62
Ross and Clark method, 165
Rudin method, 165, 182

S-methylmethionine (SMM), 129–130
Saaz hops, 86
Saccharomyces cerevisiae, 5, 101
Saccharomyces diastaticus, 132
Saccharomyces pastorianus, 5, 101
Saladin boxes, 36, 37
Sales gravity, 118
Salt, 14
Saltiness, 122
Sampling, 30
Saturated ammonium sulfate precipitation limit (SASPL) test, 146
Scarification, 35
Secondary fermentation, 115
Secondary gushing, 167–168
Sedimentation, 115, 135
Seedless hops, 84
Sensory analysis, 183–184
Separation, 69–70
Service water, 80–81
Sesquiterpenes, 89
Shelf life, 42
Shipping methods, 110
Short-chain fatty acids, 132, 207
Side chains, amino acids and, 190

Siebert model, 143–144
Sigma value test, 182
Silicate, 82, 135, 144, 145, 197
Silicone, 116
Single-use PVPP, 144
Six-roll mills, 61, 62
Six-row barleys, 26, 27, 28, 42
Skimming, 114
Skunky aroma, 169
Slide cultures, 110
Smell (*see* Aroma)
Soaps, 211
Sodium, 14, 122
Solids, 231
Solubilase, 47–48
Solubility, 54, 197–198, 206–211
Solubilization, 46–47
Sorghum, 11
Sourness, 13–14, 122
Spaerotheca macularis, 84
Spear samplers, 30
Specialty malts, 42–44
Specific gravity, 2, 7, 9–10, 114, 179
Specifications, 43–44
Spectrophotometric analysis, 85, 86
Spent grains, 4, 186
Stability, 7, 118, 137–139, 162, 180–181
Stages of commercial malting, 22–23
Stainless steel, 65, 115
Staling, 81, 137, 153–159
Standard deviation, 233
Standard malts, 41–42
Standard normal variables, 235
Standardization, 178, 235
Starch conversion, 1
Starches
  as adjunct, 68
  chemistry of, 203–204
  in endosperm, 25
  forms of in barley granules, 52–54
  gelatinization of, 50–52
  granules and, 49–50, 52–54
  visible hazes and, 141
Starchy endosperm (*see* Endosperm)
Statistics, 233–236
Stearic acid, 55, 207–208
Steeliness, 29, 32, 36
Steel's mashers, 65
Steeping, 4, 7, 22, 32–35
Steeping and germination, 1
Sterile filtration, 173

Sterilization, 4, 82, 95, 109, 118
Sterols, 102, 107, 110, 209, 210
Stevens' power law, 124
Stewing, 41
Stokes' Law, 135
Storage conditions, 7, 16–17, 31, 108, 110
Storage proteins, 54
Strain improvement, 104
Stratification, 117
Strecker degradation, 153
Strength, 9–10, 71, 113
Striking temperature, 75
Stripping, 99
Structure, 21–25, 199
Style, 9
Substrate inhibition, 196
Succinic acid, 122
Sucrose, 104, 121
Sugars (*see also Individual sugars*)
  as adjunct, 68
  alcohol and, 3
  boiling and, 97
  carbonation and, 115
  chemistry of, 201–205, 209
  flavor and, 126
  glycolysis of, 105
  mouthfeel and, 121
  nucleic acids and, 214
  sweetness and, 13, 121
  yeast and, 101, 104, 105
Sulfate, 14, 82, 122
Sulfolipids, 209
Sulfur, 104
Sulfur compounds, 16, 128–131, 152
Sulfur dioxide, 152, 156–157, 158, 159
Sun struck, 169
Super-saturation, 17
Surface tension, 17
Sweet wort
  adjuncts and, 68–69
  brew house operations and, 72–75
  calculations for, 75–76
  decoction mashing and, 66, 67
  dry milling and, 59–62
  grist cases and, 64
  infusion mashing and, 65–66
  lauter tuns and, 70–71
  mash filters and, 71–72
  mashing processes and, 64–65
  overview of, 59

  temperature-programmed mashing and, 66–68
  wet milling and, 63
  wort separation and, 69–70
Sweetness, 13, 121
Synebrychoff Company, 128
Syrups, 68

Tannic acid, 145–146, 181
Tannins, 147
Tannoids, 88, 137, 143
Tannometers, 146
Taste (*see* Flavor)
Tasters, trained, 184
Taxes, 1
Technology, 11
Teig layers, 69–70
Temperature
  alcohol production and, 126
  amylopectin degradation and, 54
  brew house operations and, 73
  chemical reactions and, 230
  cold conditioning and, 115
  conditioning and, 7
  diacetyl formation and, 128
  enzyme function and, 195
  fermentation and, 114, 115
  flavor and, 16–17, 152
  foam and, 164
  gelatinization and, 50–52, 68
  glucan content and, 147
  β-glucanases and, 47–48
  kilning and, 39
  lipid structure and, 210–211
  malt flavor and, 41
  matter phases and, 231
  packaging and, 148
  pasteurization and, 173
  solubilases and, 47–48
  specialty malts and, 42–43
  stabilization and, 144
  steeping and, 33
  stratification of, 117
  wort strength and, 75–76
Temperature-programmed mashing, 66–68
Temperature ranges, 1
Temporary hardness, 80, 82
Terpenes, 131
Tetrazolium dye staining, 28
Texture, 13, 16, 82, 121, 133
Thousand kernel weight, 28
Three-glass tests, 183
Threshold, 123–124
Thymine, 214–217
Titration, 181
Toffee flavor, 132

Tongue, 13, 15, 121
Top fermentation, 9, 102, 164
Torulaspora spp., 172
Traditional ale brewing, 7, 9
Traditional lager brewing, 7, 9
Transamination, 106
Transcription, 216–217, 226–227
Translation, 217, 226–227
Trehalose, 107
Tricarboxylic acid (TCA) cycle, 223–224
Triers, 30
Trigeminal nerve, 13, 121
Triose, 201
Tristimulus and chromaticity, 179
Trub, 5, 187
Tunnel pasteurization, 173
Tuns, 66, 70–71
Turbidity, 17, 142
Turbulence, 148
Two-row barleys, 26, 27, 28, 42
Two-way modification, 35

Units, 2
Unmalted barley, 21
Uracil, 214–217
Urethane insulants, 115

Valences, 229
Variance, 233
Vectors, 218
Ventilation, 31
Verticillium albo-atrum, 84
Verticillium wilt, 83, 84
Vessels, 34, 65, 115–118
Viability, 28, 29, 109–110, 113
Vicinal diketones (VDK) (*see also* Diacetyl), 7, 16, 115, 119, 127–128
Vienna malts, 41
Vigor, 28, 29
4-Vinylguaiacol (4-VG), 132
Viscosity, 69, 121, 133

Visible hazes, 141–142
Vitality, 110, 164
Vitamins, 18, 104, 114
Volatile compounds, 95, 99, 187–188
Volatility, 5
Volume estimation, 2, 113
Vomitoxin, 19, 167

Warm conditioning, 134
Warrior hop cones, 5
Wastewater treatment, 188
Water (*see also* Moisture content)
　as coolant, 99
　hardness and, 77, 80
　health standards and, 77, 78–79
　hydrophobicity and hydrophilicity, 193
　ions in, 81–82
　for production, cleaning, and service, 80–81
　treatment of, 82
Water hardness (*see* Hardness)
Water sensitivity, 29
Weak wort, 71, 147, 187
Weighers, 61
Weight, 2, 110
Weissbier, 11
Weizenbier, 11
Wet milling, 63, 64
Wheat, 11, 68
Wheat malts, 42
Whirlpool separators, 98, 99
Whiskey, 42
Whole-cone hops, 90
Wholesomeness, 18–20
Widgets, 165
Wild yeast, 172, 182
Wilt, 83, 84
Wine, 3
Wood chips, 135
Wort (*see also* Sweet wort), 4, 75–76, 113

Wort extenders, 97
Wort separation, 1, 69–70

Xerogels, 145
Endo-β-xylanase, 49
Xyloses, 205

Yeast
　alcohol production by, 125–126, 127
　aroma and, 16
　characterization of, 101–102
　cold break and, 99–100
　cold conditioning and, 134
　counting of, 110–111
　fermentation by, 3
　β-glucan and, 206
　handling of, 107–109
　hydrogen sulfide production by, 130–131
　image of, 6
　invisible hazes and, 141
　life cycle of, 103–104
　metabolism in, 220
　nutritional requirements of, 104–107
　pitching and, 5
　polyphenols and, 148
　recycling of, 6–7, 187
　sourness and, 13
　staling and, 157, 158
　storage and, 107, 110
　strains of, 5, 104
　structure of, 102–103
　units for, 2
　viability of, 109–110
　vitality of, 110
　wild (contaminant), 171
Yeast presses, 138–139
Yield, 30–31, 45, 118
Yorkshire square, 116

Zinc, 81, 104, 114
Zygosaccharomyces spp., 172
Zymomonas spp., 132, 171, 172